The Eternal Conflict

The complete unified theory of physics based on the universal duality and the inherent conflict between the primary force of Chaos and the primary force of Order within everyday life.

By R.A. Biers

"The Eternal Conflict: The complete unified theory of physics based on the universal duality and the inherent conflict between the primary force of Chaos and the primary force of Order within everyday life" Copyright © Library of Congress 2016, 2017, 2018 / Published 2018 by R.A. Biers. All rights reserved.

No part of this book may be used or reproduced in any manner whatsoever, including Internet usage, without written permission from R.A. Biers, except in the case of brief quotations embodied in critical articles and reviews and only when the brief quotations are used within the exact context of the overall literary work.

All interior charts and illustrations © Library of Congress 2016, 2017, 2018 by R.A. Biers.

Library of Congress © 5769428192 - 6708041704

ISBN 978-0-9993-8120-5 for eBook (2017)

ISBN 978-1-9832-7290-5 for Paperback Book (2018)

Any Internet references contained in this work were current at the time(s) of publication, but the author and publisher cannot guarantee that a specific Internet location will continue to be maintained into the future.

Force of Chaos	Force of Order
Physical Material Realm	Ethereal Energy Realm
Real Things/People	Energy, Magnetism & Electrical
Freedoms, Rights & Liberties	Control, Rules and Regulations
Anarchy Governments	Democracy Governments
Republic Governments	Oligarchy Governments
Lawful Money	Legal Tender form of Money
Lawful Laws/Common Laws	Legal Laws/Statutes
Open Systems	Closed Systems/Regulated Systems
Free Energy Systems	Limited Energy/Nonrenewable
Physical Faith-based Religions	Spiritual Belief-based Religions
God looks like physical person	God as a power or an invisible force
Statues and Rituals	No religious objects or rituals
Open Education	Closed and controlled Education
No Censorship	Censorship and Restricted Knowledge
No Assembly Restrictions	Restrictions on Assembling
Can dissent and disagree	Restrictions on Speech and Press
Hieroglyphical Languages	Alphabetical Languages
Simple Pyramids and structures	Complex Temples and structures
Ability to be Self-Sufficient	Dependency upon Government
Visible and Real	Invisible and Make-believe
Statistics	Mathematics

TABLE OF CONTENTS

Scale Chart   Page 3

Summation Quote to Ponder   Page 5

Prologue   Page 6

Disclaimer   Page 13

Chapter 1 - The Eternal Conflict in Summary   Page 15

Chapter 2 - New Unified Theory of Physics   Page 42

Chapter 3 - The Eternal Conflict in General   Page 96

Chapter 4 - The Eternal Conflict inside each Person   Page 114

Chapter 5 - Types of Religions   Page 131

Chapter 6 - Types of Governmental Systems   Page 154

Chapter 7 - Different Systems of Money   Page 178

Chapter 8 - Different Languages and Knowledge   Page 202

Chapter 9 - The two Systems of Law   Page 221

Chapter 10 - Final Summation & Message   Page 240

Chapter 11 - History of Sumer & the Language War   Page 251

Chapter 12 - History of Ancient China to Atlantis   Page 326

Chapter 13 - History of Ancient Rome to America   Page 368

Works Cited -   Page 425

SUMMATION QUOTATION TO PONDER

"The nature of energy is to be understood. The properties of space have to be intuitively derived from the available scientific data. We have to pick out a single elementary material particle to construct the material universe and to establish the unity of the material worlds.

We need to know the attributes of the single Substance, the nonmaterial <u>Akasha</u> to establish the Unity behind the diverse scientific and spiritual phenomena. We need to realize the agency of Consciousness, the principle of life and death, and the intelligence of the Universe.

We can certainly know these phenomena through a scientific approach but the science that shall answer the above questions shall not be the current science. It shall be a new science inclusive of a spiritual base. And that shall set the stage for a merger of science and spiritually". (Tewari, Paramhamsa, C.E. 1996)

PROLOGUE

The proposition presented within these pages is that there are only two primary or fundamental forces within our dualistic world. The first primary force is the <u>Physical Material</u> side, which is mostly visible, consists of all the different forms of physical matter and is called <u>Chaos</u>. Within this primary force <u>Chaos</u> includes all <u>Physical Material</u> and their sub-forces, such as volcanoes, plate tectonics, and even the erosive properties of water. The second primary force is the <u>Ethereal Energy</u> side, which is mostly invisible, consists of all the different forms of scientific energy and is called <u>Order</u>. Within this secondary force <u>Order</u> is all <u>Ethereal Energy</u> and their sub-forces, such as electricity, magnetism and radiation.

Everything within the universe from animals, dirt, planets and stars to electrical energy, magnetic currents and subatomic particles can be grouped within either the first primary force called the <u>Physical Realm</u> of <u>Chaos</u> or the second primary force called the <u>Ethereal Realm</u> of <u>Order</u>. Again, although there are many smaller forces and sub-forces, all of those can still be grouped within the two primary and fundamental forces, but it is those two main forces that actually create the dualism, which is our world.

Chaos is the ever-changing physical universe and is literally everything that is made of all the different forms of physical matter, that can mostly be seen with the naked human eye. Order is the energy grid of the entire universe that is mostly invisible and cannot be seen by the naked human eye. You can think of the force of Order as being similar to the ancient concept of the Ether, in which even Rene Descartes, the seventeenth century philosopher, thought of as a vast pool of invisible something that drags the planets along. Although this is very close to the truth, the force of Order is actually composed of all energy, like an energy grid that overlays on top of the physical Chaos and it contains all the different forms of energy. Chaos exists within this large pool of Order and the natural spinning and swirling movement of Order that actually causes the physical Chaos of planets and worlds to move, rotate and spin in very specific patterns. Suns, stars, planets and worlds do not spin, rotate and move by themselves, but rather they rotate and move because they simply float within this vast moving, spinning and rotating pool of Order.

The most important part concerning these two main forces is that they are completely opposite, but yet opposites attract and so they are attracted to each other. They are continuously <u>in opposition</u> to each other and they have always been forever in conflict. The nature of the two main forces is what is called the <u>Eternal Conflict</u>, mainly because the conflict between the two primary forces <u>Order</u> and <u>Chaos</u> has always existed in our world, everywhere within the universe, inside us on a micro level and even inside of God, if you happen to believe or have faith in a Deity. But if you don't happen to believe in God, then understand that the concept is not all that important to the subject at hand.

It is this dualistic nature of the two main opposing forces that affect not just the physics of the universe, but literally everything from religion, business, money, thoughts, laws, architecture and even explain how humans think and how we rationalize the actual world around us. These two opposing forces are also the organizational patterns and the building blocks of the universe, which cause the reoccurring patterns that exist everywhere, and include both the organized cycles of Order and the random cycles of Chaos that we all experience throughout our lives, whether we know they exist or even happen.

If you truly dive in and try to understand these two main forces, then you will also see that they help to explain a new unified theory of physics, which shall be discussed within these pages. The new unified theory of physics will describe how the universe theoretically operates, based on the universal duality of the two main primary forces that are opposing, but are also attracted to each other.

Once again - even though the two main forces are opposing, they are also blended together, again because opposites attract and together they form our understanding regarding any thought or concept within our individual life, but also within our social lives. This includes religion, politics, money, architecture, education, government, history and literally everything.

The Eternal Conflict between the force of Physical Chaos and the force of Ethereal Order tends to divide the populations into two different types of people, with

Chaos-People focusing on freedoms, liberties and rights, with a desire to govern themselves and worship who they want, within a very limited governmental structure. Order-People on the other hand tend to focus on control, laws, rules and regulations and the overall belief of society first and individuals last. They tend to downplay the rights of any individual person, due to the inherent mistakes that every physical human tends to always make, which of course impacts the collective society in many negative ways.

In our world, these two different types of people, who are molded from the blending of the two main forces, are usually completely unaware of the two forces that are guiding them. Every person that you met will usually find themselves on one or the other side of the Eternal Conflict, while at the same time believing that they are serving humanity as a whole. It is this world-wide positioning between the two opposing forces that forms the viewpoints of the two differing groups and is also what creates the great divide between Control of the Masses or the Freedom for Individuals throughout the entire planet that we call Earth.

By examining the past and present circumstances of our world, you will be able to see that The Eternal Conflict has always existed throughout history up to the present day, as it is truly the two forces that determine everything including that very history. This examination will allow you to see why the current Owners of the World have over time, separated themselves and currently exist mainly as only Order-People. As Order-People, they truly believe that they are helping the world through their attempt to form a controlled and ordered utopian society, even if that form of government causes dependence and the loss of chaotic decision-making by the freedom-loving individual people.

You will also discover that the number one major problem that humanity has always faced, and which has existed throughout history is the Lack of Balance between the two dualistic forces. This Lack of Balance usually happens but does not have to be inherent within the dual system that is called life. As long as one side, either Order or Chaos dominates and grows while the other shrinks, there is imbalance and whenever there is imbalance within the Eternal Conflict, the hopelessness of the perpetual cycles will automatically happen again and again and will always keep humanity imperfect and lacking. Once Order (control and technology) grows too strong and too large to be contained, then Chaos (freedom and populism) will rise up, which will force Order to diminish. This is one of the absolute truisms within the known world. We all know that history does repeat itself over and over, but most people do not realize that history repeats itself, only because of the imbalance that moves back-and-forth between the two main forces within the universe. This problem can be solved, but only through the understanding that both sides must have balance. Only when and if this can happen, will worldwide contentment be discovered and then occur.

Simply speaking, if you don't allow the force of Chaos and force of Order to achieve some sort of equilibrium, then they will continually exert themselves against the other, to everyone's dismay because they are actual forces.

Saint Augustine was correct when he said, "Everything in existence must co-exist in a sort of balance or symmetry". If there is no balance, then there is nothing but conflict happening, because of the lack of balance. (Great Philosophers, C.E. 2002).

DISCLAIMER

Nothing in this book should be believed. Everything in this book should be considered fictional. Flush out all the ideas and beliefs that you were taught by other individuals or select groups. Don't even believe anything written in this book but instead make up your own mind and trust nobody.

Everyone you know or will ever meet already speaks from a biased viewpoint and they will try to sway your opinion with their own agenda. Also, they will usually try to take something from you with their demands and judgments. What is written here was written to help people and not to hurt anybody. What is written here was put down to give and not to take. This book will make the soul ponder and the mind think and that can never a bad thing.

If you believe or have faith in a deity, then just keep in mind that God just desires us to learn and to grow. Exact knowledge is the only real truth and is revealed both externally around each of us and internally within us. It will shine brightly as it is revealed, not necessarily through our words, but mainly through our actions or fruits.

This book is a work of fiction, written for entertainment purposes only. Nothing here should be considered financial, legal, lawful or medical advice. Names, characters, places and incidents either are the product of the author's imagination or are being used fictitiously. Any resemblance to actual people (living or dead), deities (living or dead), events or locales is entirely coincidental.

CHAPTER 1 - THE ETERNAL CONFLICT IN SUMMARY

Trying to understand our Dualistic World

We live in a dualistic world - a world where almost everything is described, rationalized and eventually understood within the context of only two things or two ideas. Where there is an **up** in our world, there must of course be a **down**. Where there is an **inside**, there must of course be an **outside**. Where there is **light**, there must also be **darkness**. This is because our physical brains and our ethereal thoughts within those brains, naturally wrap themselves around dual patterns and it is within these dual patterns that our human brains understand things easily and quickly.

When somebody is accused of a crime, most people want to know whether the accused is **guilty** or are they **innocent**, sometimes even ignoring the facts of the case, without even wondering if the accusations point in a different direction. To keep things simple for our dual-minded brains, we may even jump to conclusions and even ignore facts that prevent us from making that quick judgment.

Although the world tends to be _gray_ in color with lots of ambiguity, humans want and desire only the _black_ or the _white_ to all situations. This harsh dualism of absolutes has always been deep inside us, both individually but also socially within different groups and it is this dualism that tends to mold the life that we know both individually and collectively.

Many great writers and artists, throughout history, have reminded us that reality has never been about absolutes, such as _white_ or _black_, but rather life is mostly made up of the _gray_ areas in between those. Absolutes are actually very rare, yet we crave them and want them. We live in a dualistic universe, dualistic world, dualistic society and we are even dualistic individuals that try to understand life in two patterns and those two patterns are constantly reinforced within us, as we go through our daily lives compartmentalizing everything as simply _yes_ or _no_.

Why does this happen? Many people believe that the answer is perceptional or behavioral, but the truth is not either of those things, but instead it is the reality that we live in a dualistic world, because literally everything within it and even deep within ourselves is made up of only two main forces and those two very different forces literally create and form our very ideas and thoughts. If you doubt that fact, then keep in mind that everything today (machines, computers, cash registers, etc.) run off electronics that are programmed with Boolean logic, the simplest dualistic language that only has two values, <u>1</u> or <u>0</u> and those two values change through an electronics switch making those two values switch either <u>On</u> or <u>Off</u>.

George Boole, the philosopher and inventor of Boolean mathematics and Boolean logic, back in the beginning of the nineteenth century, proposed that the entire universe operated according to very basic rules and if everything in the universe operated according to those rules, then even the way people believe or even the way that the mind of God works would be based on those same very simple rules. Rules which George Boole then converted down, within his new mathematics, to be based on two logical values. As he stated, "All sciences consist of general truths, but of those truths some only are primary and fundamental, others are secondary and derived. The laws of elliptic motion, discovered by Kepler, are general truths in astronomy, but they are not its fundamental (primary) truths. An almost boundless diversity of theorems, which are known, and an infinite possibility of others, as yet unknown, rest together upon the foundation of a few simple axioms." (Boole, George, C.E. 1854).

This idea from George Boole of very simple dualistic rules has been proven as absolute truth and this truth can be visually seen and even understood today, within computers, electronics and even the emerging artificial intelligence sciences. I would add that this now proven postulate of George Boole, which is the basic understanding of the universe, shown through computers and computer programming today, complements the writings in this book. The new postulate which says that this universal system operates, because there are only two forces within the universe and they are at work in our dualistic world and those are the force of <u>Order</u> and the force of <u>Chaos</u>, which we shall now discuss in greater detail.

The first of two primary forces is Chaos or the Physical World

The first primary force is <u>Physical Matter</u> and within this classification is everything material such as dirt, rock, water, flesh, wood, etc. If it is made of matter and is usually visual to the naked human eye, then it can be found within the classification of Physical Matter, which is the force called Chaos. There are a few notable exceptions to the understanding that physical matter is usually visual and those are the physical gases, such as hydrogen, oxygen or nitrogen to give you some examples. These gases are still physical matter but since they are in the form of a gas, then they seem to be invisible to the human eye, unless they are of course cooled down or heated up.

There are many who would argue that the Physical World, which is made of matter, is not actually a force at all. But that is not true and if you just sit and observe your surroundings, looking at the trees, the oceans, the mountains, the animals and the people that make up the Physical World from an honest standpoint, then you will see a moving coherent force at work, that today is sometimes called Mother Earth but in the ancient past was the force called Gaia. As the Earth moves and its tectonic plates shift to create earthquakes, the resulting physical sub-forces become tremendous. Where the molten lava of volcanoes will erupt and literally change the landscape around it, you can see and measure the resulting destruction. In our observed reality, even the smallest creation of asexual organisms or the gentle births of animals are not only beautiful but a measurable force and power.

Everything within the primary force of Chaos or Physical Matter does not follow precise mathematic formulas, but instead they follow only statistics or predictable patterns and trends. Chaos as nature follows probability theory, which can be measured to find an average length, average height, average weight, etc. but those are never exact and perfect, they are rather just a series of measurements that always fit within a statistical Normal Distribution Curve. This is unlike the exact formulas and equations that are used within the opposing force called Order.

Physical Matter is demonstrated as the force of Chaos and is reinforced within that primary force through the understanding that everything that is made of Physical Matter has a Chaotic Nature. All Physical Matter constantly changes, evolves, devolves, breaks down, gets old, is recycled or simply dies. Even after physical death, the matter will continue to change into something else. The chaotic nature of physical matter is why statisticians, historians and futurists always struggle to study or predict the behavior of anything made of Physical Matter, such as human behavior or the patterns of societies and communities. The reason is that human behavior is always very chaotic and unpredictable, and it is the same for any physical pattern or physical study, such as earthquakes, storms or when the next volcano will erupt and where? Physical Matter as the force of Chaos is exactly chaotic.

In general, the force of Chaos or the Physical World can never be measured in absolutes or with absolute accuracy. That is due to the inherent randomness that is within the force of Chaos. Within the study of probability and statistics, the inherent randomness within the force of Chaos is called normal variation or internal deviation.

The same D.N.A. from a man and a woman will create many varied and different offspring, all having varying traits, such as their height or their size. The Physical World made of matter can only be viewed by its natural variation or inherent randomness and can only be measured through probability and statistical methods and those can never be exact. Measured Chaos will always include tolerances, deviations and spans. What is interesting however, is that if you have a large or big enough sample of whatever physical attribute or physical matter that you are studying, then it will result in the force <u>Chaos</u> defining the span or scope for that situation. For example, if you measure the length of grass blades in an open field and if you gather enough samples, then you can use the <u>Normal Distribution</u> aspects, within the science of statistics, to define how tall the tallest blades of grass should be or how short the shortest blades of grass should be, while at the same time determing how many of each height will be found there, even before you find and measure them all. This is called <u>Adequate Sampling Size</u> and is just another example which demonstrates that the force of <u>Chaos</u> as <u>Physical Matter</u> is an actual active force within the universe.

The force of <u>Chaos</u> also includes the physical chemicals and hormones that are both outside and within the human body. It is these physical chemicals and physical hormones that directly cause the chaotic behavior patterns that happen within animals and people, because their emotions and emotional situations are nothing more than their internal chemical response to either internal or external stimulus. Emotions, such as bursts of anger, mood swings or even sudden tears of joy are all very chaotic behaviors, due to the uncontrolled or unplanned nature of those behaviors as they are considered to be a part of the force called Chaos.

Chaos or Physical Matter is the direct opposite of the other primary and fundamental force called Order and is the very reason that the force of Chaos does not like to be controlled, organized or even ordered. There are times, when most humans prefer and even behave in ways that tend to pick individual choice rather than group thought. Going even further, some people even demand individual freedoms, rights and liberties. In fact, this idea has been called many things, such as Free Will, but from the perception of the force of Order should be called Chaotic Decision Making. When a society or government is developed around the idea that each individual citizen should be free to choose their own path and make their own decisions, it is important to understand that people are fallible and will sometimes make terrible decisions and bad choices, especially when put into large groups. This is mainly because decisions that can be made by physical people are definitely a by-product of the force of Chaos.

The second of two primary forces is Order or Ethereal Energy

Again, there are only two primary forces in the universe and the second primary or fundamental force is called <u>Order</u> and consists of all the <u>Ethereal Energy</u>, which is almost all the energy systems everywhere and they are mostly invisible such as electricity, radiation, visual light, magnetism, and photons. If it is invisible to the naked human eye and could cause damage or changes to anything made of physical matter, because it is its direct opposite force, then it can be found within the classification of <u>Ethereal Energy</u>, which again is the primary force that we call <u>Order</u>.

The easiest way to understand this second force called <u>Order</u> or <u>Ethereal Energy</u> is to look around you at all the seemingly empty space around your body. This invisible area may look like empty space, but it is not. Any invisible looking space anywhere within the universe is filled with massive amounts of <u>Ethereal Energy</u> such as static electricity potential, radio waves and magnetic currents, while the seemingly empty space just around your body is also filled with radio-wave frequencies, television signals, satellite information and even cell phone signals. You can prove this easy enough, because if you have a radio, television or a cell phone operating, then you can simply just raise up the antenna and then turn its dial to tune into all of the massive amounts of <u>Ethereal Energy</u> that is literally all around you.

In addition to that, the seemingly empty space is also filled with so-called visual light that strangely is invisible but exists to make physical things visible to the human eye. When a room is completely dark and then you flip on a light switch, the room fills up with invisible visual light. You can't see the visual light, but the energy of the visual light does allow you to see everything else in the room. This is because the visual light now bounces off everything and then returns to your human eye, which then allows you to see the <u>Physical World</u> around you.

Visual light does not however, allow you to see the other half of your world, which is the <u>Order</u> side that is made up of <u>Ethereal Energy</u> and that is even though the other side exists right around you and in front of you, mainly because it is invisible to your limited human eyeball. The main reason that all the sub-forces of <u>Order</u> are usually invisible to the human eye has to do with internal frequency and internal speed within the sub-forces that are a part of the force of Order and which will be discussed in greater detail later. The most important part for now is that Physical Matter or Chaos can usually be seen and visible, whereas Ethereal Energy or Order is usually invisible to the naked eye, although it is everywhere, wrapped around and controlling Chaos.

This is a critical idea that you must try to understand, so it will be restated - all the seemingly empty space around you, throughout the world and even in the seemingly void of outer space is not empty and never has been, but rather is actually completely full of the second primary force called Ethereal Energy that pushes down and against all physical material and all of the sub-parts of the first primary force called Physical Matter. It is this interaction between both of the two forces that actually gives Physical Matter all of its physical form and shape, plus it helps direct its position and even the movement of the Physical Matter that exists within it.

Ethereal Energy as the force of Order is reinforced through the understanding that everything made of Ethereal Energy has a very Ordered Nature. For example, electricity and magnetism are so ordered that we have numerous exact mathematical equations and precise formulas that describe their behavior and patterns. In addition to that, the actual behavior of any type of Ethereal Energy has never been random or chaotic but instead is incredibly calculated and ordered, making its predictability perfect and exact.

This force called Order or Ethereal Energy is all the energy systems within the universe and as mentioned, the most important thing to remember is that the force of Order can be precisely calculated with perfect and unchanging mathematical values and equations. Order is exact and precise, unlike the other force called Chaos, which can only be guessed at using statistics and probability methods that determine averages that will have tolerances.

Keep in mind that the force of Order as Ethereal Energy is many things and may or may not include what humans call the mystical or spiritual realm of spirits, ghosts and other invisible energy creatures that we may not even be aware of because of their inherent invisibility, but this force definitely includes our thoughts, ideas, mathematics and even information. I find it interesting that this concept of <u>Information as Pure Energy</u> has been around for a long time, but it is much easier for us today to understand and grasp in our modern world of computers with electrical memory and electrical information storage systems, then it would have been in the much distant past. The concept of technology as a growth of ideas and information can also fit within the force of <u>Order</u>, because as electrical technology, electrical controls and electronic information storage systems grow, so does the force of <u>Order</u> as its controlling aspect. If you can understand this concept, then you that would also have a better grasp of the idea that past civilizations could have developed high technology, like us today, but then as <u>Order</u> becomes overwhelming due to technology, <u>Chaos</u> will also automatically intercede to try and achieve balance, which could cause any society to instantly deteriorate.

The two primary and fundamental forces, <u>Chaos</u> and <u>Order</u> are exactly opposites from each other and therefore act in many ways as complete opposites, but they also are incredibly attracted to each other, since opposites always attract. This is what gives the universe the inherent need for balance and harmony, since both forces are at work within the universe opposing each other but attracted to each other. This is an incredibly important fact, because although both forces are in direct opposition to each other, they also have an attraction which causes them to blend together and this blending together actually causes additional conflicts to occur between each other. To stop the conflicts, there must be a balance. Anytime there is peace, there is an equilibrium between the two primary forces.

A perfect example of the opposing nature of the two forces is the change that each of them experience when exposed to each other. The behavior and patterns of <u>Physical Matter</u> will change and devolve slowly as a normal <u>Chaotic</u> part of that force, but they will also change instantly every time they contact or experience direct <u>Ethereal Energy</u> or <u>Order</u>. For example, electricity or radiation will alter and can even damage or destroy physical human flesh upon contact.

On the other hand, the behavior and patterns of <u>Ethereal Energy</u> or <u>Order</u> can only be manipulated by using <u>Physical Matter</u> or <u>Chaos</u>. Its internal <u>Ordered Nature</u> refuses to change or be altered by <u>Chaos</u> unless forced to and usually through the manipulation of <u>Chaos</u>. For example, visual light can be bent with physical water or broken apart with the use of a physical crystal prism. Electricity naturally is attracted to physical matter, specifically physical ground, but the electricity can also be manipulated on a physical circuit board by forcing the electricity to move around physical copper strips and through the various physical circuits. The electricity does not like being manipulated and will short-circuit the path or the flow of electricity around the circuit board whenever it can, but the reality is that both forces are completely in opposition to each other, which is the very nature of the <u>Eternal Conflict</u> between the force of <u>Chaos</u> and the force of <u>Order</u>.

<u>Our dualistic world made up of Chaos and Order</u>

The idea of the dualistic universe being regulated by the primary force of <u>Chaos</u> and the primary force of <u>Order</u> is a very ancient idea and almost all of the religions, both ancient and modern, refer to our dualistic universe as being made of two sides. The main difference today is the <u>development of the idea into a new kind of physics</u> that is simple and straightforward.

In the past, many different religions would use different words such as the Physical Earth to describe

Chaos or the Spiritual Heavens to describe Order. They would also maintain the idea that a Physical Realm made of Chaos would be full of chaotic events that create sin and confusion, while the Heavens would be seen as being composed of the Ethereal Realm made of Order with many absolute rules, regulations and laws, especially the laws of a judging God.

Other ancient writings reveal additional truths when they say that the two forces that are in existence are in conflict or opposition to each other. For example, there is much discussion within numerous religions that the main conflict of humanity is between the Flesh of the Human Body that wages war against the Essence of the Ethereal or Spiritual Soul. Some religions will even go as far as to state that to save the Essence of the Ethereal Soul, each person must be willing to give up the Flesh of the Human Body through physical pain and material suffering. In fact, these two conflicting sides of our dual reality have been shown in numerous religious symbols, with two such examples being that of the Asian Yin and Yang symbol or the many different images of two-sided or two-faced Deity, such as the Roman God called Janus. One of the reasons that the Eternal Conflict is described through ancient religions is because it creates internally the very dualistic reasons that everybody thinks the way that they do, and it is also why each person's sees the world through their own version of reality either <u>Order-Based</u> or <u>Chaos-Based</u>. The <u>Eternal Conflict</u> as a new theory of physics is the truest concept that most philosophers and scholars have used to both explain and teach people throughout history.

By viewing everything in the universe through a relationship between the <u>only two primary forces</u> that exist and their inherent <u>Eternal Conflict</u>, you will easily see and understand why all of humanity sees all of life as a duality, because everything that exists has two dual forces acting upon it and it is this interaction between the forces of <u>Chaos</u> and <u>Order</u> that cause life to exist, as the duality that we all try to understand.

<u>Chaos and Order has a need for balance</u>

As mentioned earlier, when the two primary forces, <u>Chaos</u> and <u>Order</u> become equally blended or mixed together evenly, then they also begin to work together seemingly in harmony, but only whenever a balance can occur between them both. It is this balance, where equal measures of each primary force can be found that should be the goal of all things. It is this need for balance that also creates the attraction between the two opposite forces within the universe, even though they have been opposed to each other eternally. The truth is that if we don't allow both <u>Chaos</u> and <u>Order</u>, some freedom and some control to both exist in equal measure or in some sort of balance or equilibrium, then the two forces will try and force a balance themselves anyways. This will usually happen automatically to the dismay of those around them, because both are actual forces that will exert themselves.

To give you an example from political history, a new form of government starts from scratch as a free <u>Republic</u> (Chaos) form of government for its chaotic freedom loving people, but this creates a government with no balance. Then when multiple bad chaotic decisions start to be made by citizens or by the government, then rules and regulations (Order) start to be added to try and help the people until the form of government changes to a <u>Democracy</u>. But as more rules and regulations are added, to try and achieve balance, them the force of Order starts to overwhelm until a dictator arises and takes control of the populace. It is at this point where the government usually then descends into an <u>Oligarchy</u>, which usually continues until a revolution occurs (Chaos) once the people feel overburdened and totally controlled by their own government.

This <u>Eternal Conflict</u> political example can be seen from within a historical basis concerning the Grecian Republic, Roman Republic or even the current American Republic and demonstrates historically that the <u>Force of Chaos</u> exerted itself from within, because the form of government was unbalanced, at least at the beginning with very little <u>Order</u>, so the <u>Force of Order</u> slowly exerts itself and create rules and regulations, supposedly for the benefit of the people. In reality though, it is the <u>force of Order</u> trying to create balance. Eventually the force of <u>Order</u> will become too powerful and take control of the people lives and at that point, there is now very little <u>Chaos</u> left, so then the <u>Force of Chaos</u> will rise up, such as a revolution of the people or something else, to try once again to rebalance the forces of the universe within this small example. If you can understand this example from history, you should be able to see why a balance is necessary and will be purposefully created between <u>Chaos</u> and <u>Order</u>. In this example, from within their respective political realms, history is bound to repeat itself over and over again, unless something changes to achieve true balance.

As mentioned earlier, two of the universal principles within the <u>Eternal Conflict</u> are that of <u>Opposites Attract</u> and <u>Similar Things Repeal</u> and these principles can be easily demonstrated to be a part of reality with magnets, whereas the positive charge of one magnet repeals the positive charge of another magnet, but the positive charge of one magnet attracts and joins together with the negative charge of another magnet quickly and easily. Again, although the magnetic charges are in direct opposition to each other, there will always be an attraction due to the need for balance. Again, in all things, it is when things go out of balance that problems surface.

To see another perfect example of equal balance and harmony between <u>Chaos</u> and <u>Order</u>, then you only need to look within your own body. The human body is just like everything else within the universe as it exists as a duality. The <u>Physical Matter</u> is our physical body made of flesh, blood, water and other physical elements but it is also energized, from within, by <u>Ethereal Energy</u> or electricity that runs throughout our brain and down every nerve and into every cell of the body. It is the blending of these two forces, in equal balance, that allows us to have a heartbeat and muscle movement and even to receive thoughts and ideas. Although both sides work together well when they are in balance, there can be major problems, such as illness and disease when they don't. Also, humans in general usually struggle daily because of the opposition between the laziness of the physical body fighting against the consistent and constant drive of the human ethereal spirit, consisting of the electricity running throughout our body that keeps our heart beating and keeps us all alive.

For a larger perspective of equal balance and harmony, we can look to outer space that is incorrectly described by some scientists as just a vacuum of nothing. Outer space consists of <u>Physical Matter</u> in the form of planets, comets and other planetary bodies, plus all the matter that is in the form of gases, such as nitrogen or hydrogen. Outer Space is also full of <u>Ethereal Energy</u> such as electricity, photons, plasma and magnetism.

There is much debate today concerning this idea, but no matter the disagreement, the fact is that outer space may look like a vast emptiness, but it is actually full of <u>Ethereal Energy</u> as we can receive radio signals from it and we can even hear energy signatures coming from outer space. We also know that magnetism exists and is absolutely created when electricity moves. Since there is lots of magnetism within our universe, it would stand to reason that the universe is electrical and moving, since that is where the magnetism comes from. We also know that the primary force of <u>Order</u> is mostly invisible, but when you look upward to the heavens, you see a universe that is definitely not empty but rather it is full of invisible energy bubbling, spinning and churning. Again, this invisible force called <u>Ethereal Energy</u>, that is everywhere within our universe, is made up of numerous forms of energy that in ancient times was called <u>Ether</u> and which today can be easily detected as a background noisy hiss of energy.

I will acknowledge that the idea of an <u>Electric Universe</u> is very controversial and is usually quickly dismissed by many physicists and scientists today, even though there is much evidence for its existence. The idea for this alternative theory of physics was actually first mentioned in the year C.E. 1883 (Torpedo, C.E. 1883) and has been consistently revised over the last century and a half, with the latest revelations coming from the book entitled <u>The Electric Universe</u>, (Körtvélyessy, László C.E. 1998) and also one book of a series of books called the <u>Thunderbolts of the Gods, A Radical Reinterpretation</u>

of Human History and the Evolution of the Solar System, (Talbott, David; Thornhill, Wallace, C.E. 2005). In addition, there is Charles Bruce, Astronomer and Physicist (Doctorate of Science), Edinburgh University, who throughout the C.E. 1960s and 1970s in America successfully published over one hundred papers proving that most of the cosmological effects and phenomena that can be found and observed within and throughout the known universe has an electrical basis and is definitely electrical in nature. (Bruce, Charles E.R., C.E. 2018).

In his book, The Electric Universe, Doctor Körtvélyessy, who is a worldwide specialist in electricity and thermocouples, "discovered that the Universe contains many unsolved mysteries for which scientists create things like dark matter or magnetic generators ... (while) simple calculations reveal that there is a very simple solution (that is electrical) which unlocks almost all modern mysteries". (Körtvélyessy, László C.E. 2018).

Some of the evidence for an Electric Universe is a faster rotation of our Sun at the equator (without an equatorial bulge) verses a much slowly speed at its poles, proving that the Sun is using up external energy (invisible plasma) at its equator and not using fission power from within. There are also two bands of fast moving electrical winds encircling the Sun, which is common to all planets, including the Earth. (Acheson, Mel, C.E. 2013).

You can also see this <u>Ethereal Energy</u> as it blends with the <u>Physical Matter</u>, such as when comets move and glow like neon signs as they move throughout the universe, or when plasma forms colorful distant nebulas and star systems. The universe gives off massive amounts of <u>Ethereal Energy</u> in the form of magnetism, electricity, radiation and radio waves and it is important to note once again, that these can only exist with the movement of massive amounts of invisible electricity or electrical plasma. The magnetic characteristics of an electrical motor exist only because of the electricity that moves through the wires, once it is turned on. In much the same way, the magnetic areas of Outer Space and within the universe exist only because of the electricity that runs through it, creating the magnetism and other energies. Again, it is a blending and balancing of both <u>Chaos</u> and <u>Order</u> that complement our entire universe.

"Lightning is electric - like a welding arc. The Northern Lights are electric - like a neon sign. A sunny day is electric. We work and play in an electric field. We don't think about it because we're used to thinking only gravity is there. (However) space probes find electric features in comets, planets, stars and galaxies. X-ray and radio telescopes find electric features connecting stars and galaxies. The Electric Universe is a way to being thinking about all that electricity in the cosmos". (Acheson, Mel, C.E. 2018).

If you can picture and understand that the force of <u>Order</u> as <u>Ethereal Energy</u> is invisible and everywhere within the known universe, then if you can also picture and understand that the force of <u>Chaos</u> as <u>Physical Matter</u> may be visible but is not the only force in the known universe, then if you can also picture the ever-moving, spinning and charged force of <u>Order</u> overlapped around <u>Chaos</u>, around all matter, only then can you begin to see that all matter gets its movement, rotation and even shape from the constant presence of this energy. These two forces make up everything and are related to each other as the actual duality that we live in. It is simply the <u>Physical Realm</u> of visible matter and the <u>Ethereal</u>

<u>Realm</u> of invisible energy that is blended together, and which form the duality that we know.

With that in mind, let's jump in and start reviewing this new unified theory of physics in a broad sort of way.

CHAPTER 2 - NEW UNIFIED THEORY OF PHYSICS

The Eternal Conflict fits within Occam's razor

Scientists and Physicists for centuries have worked hard at trying to define the universe, with one of the newest goals being that of describing everything within one theory or one equation, which would be called a <u>Unified Theory of Physics</u>. Although I know that I am know being controversial, I do truly believe that there is one <u>Unified Theory of Physics</u> and that it can be explained easily, but only within the context of the <u>Eternal Conflict</u> and the two main forces called <u>Chaos</u> and <u>Order</u>. It is the very concept that George Boole searched for in the early part of the nineteenth century. I also know that this statement is even more controversial because I am not a physicist or a scientist, but I do understand that physics today, in my opinion, has moved away from simplistic ideas and simple observation into a realm of extreme complexity.

Physicists in the past had many simple ideas and thoughts about the world around us that were based on what they observed, while physicists today now develop strange theories about quarks, dark matter, strings, time travel and that Outer Space supposedly warps and curves, all of which are far-reaching but also extremely intricate. I assume that physics today has moved to a place in a large forest, where they can no longer see the entire forest because of the biggest tree that is directly in front of them.

Author Mark McCutcheon within his book <u>The Final Theory - Rethinking our Scientific Legacy</u> stated that "Science today contains theories that cover every known observation, collectively known as <u>Standard Theory</u>. It may seem that <u>Standard Theory</u> provides us with a fairly comprehensive scientific understanding of our universe. But is this really the case? How much do we truly understand gravity, for example? Does it really make sense that a force holds objects to the ground, and moons and planets in orbit, all with no known power source? Do we truly understand light? Today we have settled on a belief that somehow light is *both* a wave *and* a particle - sometimes manifesting as one and sometimes as the other, depending on the situation or experiment ... (this theory is) described by its very creators and practitioners as bizarre and paradoxical. Do we truly understand magnetism? Is it reasonable that an apparently *endless* force from within magnets will continually battle any external power source ... (while) there is *no identifiable power source at all* within these magnets to support the supposedly endless force that occurs from within? Do we even know what magnetic fields are, or have we simply discovered how to create then and learned to model their behavior with equations? Are we confusing practical know-how and abstract models with true knowledge and understanding? Science has managed to *model* our observations rather well ... but we have little clear physical explanations for why they behave as they do". (McCutcheon, Mark, C.E. 2010)

Author Pascal-Emmanuel Gobry in his article entitled Big Science is Broken, published in THE WEEK online magazine wrote the following, "The current (scientific) system isn't just showing cracks, but is actually broken, and in need of major reform. There is very good reason to believe that much scientific research published today is false". (Gobry, Pascal-Emmanuel, C.E. 2016). In my opinion, what is being suggested by these amazing writers is that today, physicists and scientists are not only reaching for strange ideas and weird theories, but they are also faking some of the evidence.

To give you a modern example where science today sees only the big tree and not the forest, there is currently a theory of physics called the <u>Superstring Theory</u> and without going into too much detail, it supposedly merges the general relativity theories of Albert Einstein with current various ideas about quantum mechanics. Except that within this theory there simply ends up being a lot more questions then there are answers. "Superstring theory ... comes with many side-effects which all too often go unnoticed. To begin with, the *super* isn't there to emphasize the theory is awesome, but to indicate it's supersymmetric ... symmetry that postulates all particles of the standard model have a partner particle ... (but) these partner particles were not found". (Hossenfelder, Sabine, C.E. 2018).

This is simply modern science searching in the dark but also I would completely disagree with the final sentence where it stated that although everything is dualistic, each particle has a partner particle, if you introduce the new unified theory of the <u>Eternal Conflict</u>, then you would easily find that all the particles of <u>Chaos</u> called <u>Matter</u> do have a partner particle, with is the opposite particles of <u>Order</u> called <u>Energy</u>, it is just that the partner particles of Order are invisible to the naked human eye.

With that being said, I would simply like to paraphrase Ockham's razor, which is a major problem-solving principle that has been attributed to a Franciscan friar called William of Ockham, around the beginning of the 14th century where he said that, "whatever hypothesis or idea has the fewest assumptions or fewest parts should be selected. Life is actually very simple, and we should not try to impose complexity upon such simple systems". (William of Ockham, C.E. 14th Century)

It is my personal opinion that the Eternal Conflict is the simplest of such ideas, plus the fact that I also believe that as a theory, it can be used to easily explain and understand all of the physics of our world, ourselves and that of our entire universe. I remember studying physics as a child and building rockets to understand momentum and thrust. To me, studying physics was not that complex and the ideas within physics should also not be that complex. This is absolutely true within the idea of the Eternal Conflict, where there are just two primary forces within the universe, at work against each other, opposing but yet attracted to each other.

Five-Point Summary of the Eternal Conflict

Before moving on and giving more details and additional understanding to the Eternal Conflict theory, it is critical that we stop for a second to once again summarize what is the Eternal Conflict by listing it as simply five points.

THE FIRST POINT - The first force within the Eternal Conflict is that of Physical Matter or Chaos and it is real and is actively at work within the universe. Strangely, since this force is usually visible, it is sometimes difficult to see it as an active force. Some people may look at a rock and they do not see a force at work, but if the rock was a lump of coal or a piece of phosphorous, then maybe it would be easier to understand it as a force. You may even have to look at Physical Matter through its interactions with other forms of matter to see it as a true active force. For example, you take two simple physical chemicals and mix them together and you can get a chemical reaction, heat or even an explosion, all from a simple rock that is part of the first force called Chaos.

THE SECOND POINT - The second force within the Eternal Conflict is that of Ethereal Energy or Order and it is real and is actively at work in the universe. Strangely, since this force is invisible but contains most of the sub-forms of energy such as electricity, radiation or magnetism, it is easier for the average person to see Order or Ethereal Energy as a force. This is because this force opposes and can be used to change or destroy physical matter and each person can physically see those visual changes.

THE THIRD POINT - The two main forces are a part of everything and literally make up everything that exists. They are at work within each person, each animal, each planet and each star, the entire universe and every day, they influence our bodies and our minds and help us to form our personal beliefs, individually and socially.

THE FOURTH POINT - Both of these main forces are opposed to each other, but are also attracted to each other, which causes the strive for balance between them. This point is critical to our understanding of not only the new physics, but to understand social structure, political understanding, religious beliefs and the ways that money, trading and banking systems are run.

THE FIFTH POINT - Since there is such an automatic need for balance between the two forces, then once either force becomes too large in any of its spheres of influence, then the other will rise up to retain balance. This is true even though many historical events that have occurred, where it was the force of Chaos or Order that arose, are usually attributed to a specific person, place or event. I would also add that any out-of-balance situation that exists within any of the various spheres of life, whether they be political, social, monetary, religious, etc. will automatically cause major cultural and/or social transitions. The person or place may be consequential, but the adjustment toward Chaos or Order is automatic and will happen no matter the person or place. It is the very reason why a shift specifically toward the force of Chaos is usually never predicted or even considered real by powerful people, until after it has already happened.

With this five-point summary in mind, let's now begin to go further and explore the new unified theory of physics, based on the Eternal Conflict.

New Unified Theory of Physics - Internal Vibration

As described earlier, there have been many attempts by physicists in the past and even today to try and understand the world through what is called a unified theory of physics, which is defined as one theory that would explain everything and every relationship between everything in the universe with one condensed statement, formula, equation or theory.

In general, the dilemma that confronts physics and physicists is that they are trying to incorrectly understand Physical Matter and Ethereal Energy as two completely separate things, that only have a minor causal relationship, instead of the reality that they are actually the main two forces that completely envelop the other and impact each other. The truth is that you cannot understand one force without understanding the influence that the other force has on it.

With that in mind, the first part of the new unified theory of physics called the <u>Eternal Conflict</u> starts from within each of the two forces at a subatomic level.

<u>POSTULATE 1</u> - All of the subatomic particles of either the force of <u>Chaos</u> or the force of <u>Order</u> are identical in nature. What makes every atom, cell or molecule unique may be its attributes, but primarily it is the internal speed, frequency, vibration, spin, twist and torsion of those subatomic particles from within, that create the uniqueness and identity of each item. The differences in the internal speed and spin are actually what makes or creates the item and the classifications within each sub-particle. Together those different internal speeds and spins create the exact nature of <u>Chaos</u> and the exact nature of <u>Order</u>.

To look at it a different way, every piece of reality, in the duality of our universe, can be broken down into the same extremely small subatomic particles and there are only a finite number of different particles that make up everything within our universe and within both forces. This idea may seem strange, but everything in the universe is built upon by the same building blocks and even the prevalent theory about atoms today is similar in concept. Scientists today say that all atoms are constructed of protons, neutrons and electrons. This current theory about atoms also states that you can just adding more building blocks, more protons, neutrons or electrons than you could literally create everything else. According to today's theory, an atom of the gas hydrogen supposedly has only one electron, but if you could add just one more electron to its atomic structure, then magically you will now no longer have a hydrogen atom but instead you now have the gas helium, which is kind of weird to explain and almost makes no sense.

I say this theory almost makes no sense, because after having a conversation with my brother Richard Biers, he commented that we are taught that supposedly protons and neutrons together make up the center or nucleus of every atom and they have a positive charge, while there is at least one electron spinning above and around the center of the atomic nucleus, even though this electron has a negative charge. This theory is believed and even taught to others, even though we all know that opposites attract and so any negatively charged particle (the electron) would not spin and float above the nucleus, but it would instead speed toward the positive center of the atom. That is because positive and negative charges are attracted to each other. Why would the negatively charged electron stay spinning above the atom instead of moving down and connecting with the opposite positive charged nucleus?

This is especially true for gases that are cooled down and start to become liquid, such as steam becoming water or the gas nitrogen becoming liquid nitrogen. When this happens, we know that lowering the temperature of those gases, supposedly causes the electrons to literally slow down during their rotation, which should in fact, help increase their attraction to the positively charged center. But again, that never happens. Lastly, if on the outside of every atom there is at least one negatively charged electron rotating around it, then why would those negative electrons on one atom then supposedly move closer toward other negative electrons on other atoms, forming molecules? These are questions that seem to have no answer, at least for now. See <u>The Electron Myth</u> for more details. (Biers, Richard Lee, C.E. 2011)

Again, the author Mark McCutcheon goes further when he states that science today "cannot even be certain we have properly characterized the fundamental forces of nature. If, for example, our theory of <u>Electric Charge</u> is an improper model of the true underlying principle behind many of our observations, then our current model of proton behavior as positively charged particles that always repel each other may not be an accurate description of the nucleus of an atom ... the further concept of a <u>Strong Nuclear Force</u> keeping the nucleus from flying apart would be a completely unnecessary fabrication, and our attempts to find a unifying theory would be based in part on forces that are misunderstood or ... based on such flawed assumptions from the start". (McCutcheon, Mark, C.E. 2010)

The Eternal Conflict as a new unified theory of physics holds the belief that the sub-atomic particles within atoms do not have any charge at all or if there is a charge, it constantly varies or changes. Instead of believing in a flawed assumption about inherently charged particles, maybe instead all of the extremely small subatomic particles have something else that identifies them, which is that they all have an internal frequency or internal vibration and it is this internal frequency that causes that entire atom or molecule to change into its completed entity. Also note that as the internal vibration, rotation and spin, which occurs at extremely fast speeds happen, then this also means that every subatomic particle within the entire universe is always under torsion. They are vibrating and spinning within a rotating system, much like our solar system, but at a subatomic level. I believe that no subatomic particles ever move straight but rather, they create a small vortex under torsion during its movement cycle and this rotating, twisting and vibrating nature of everything is what defines both types of forces, either Chaos or Order.

It doesn't matter whether it's a particle of Chaos or a particle of Order, because all particles of both primary forces would spin as they vibrate, move, and create torsion at a subatomic level. I am also suggesting that it is their internal speed or internal vibration, within each subatomic particle, that defines each entity and its sub-atomic grouping. Remember that they both use the same subatomic particles, but it is their internal speed or internal vibration that forms them into whatever element they become. Also, after they reach whatever internal speed and vibration that is specific to the element that they are, then you must ask the question of what determines whether they are visible or invisible to the naked eye? I would suggest that the dividing line between particles that are either from the visible

Physical Matter or invisible Ethereal Energy is the same rotating and vibrating internal speed of the subatomic particles that make it up and that are within the grouping of each entity.

To explain in greater detail - if the subatomic particles spin, vibrate and rotate at a speed slower than the speed of invisible visual light (C), then we can generically put them into the classification of Physical Matter as they will reflect visual light and can be seen by the human eye. The internal speed of the particles of Physical Matter are much slower than the internal speed of the particles of Ethereal Energy and it is their own slower internal speed and vibration that allows the subatomic particles to reflect visual light, become visible and seem solid to the human eye.

Much like a toy top that is at first spun very fast, it will stay upright and start out as a very stable rotating object. However, when its spin starts to slow down, it then becomes unstable and starts to wobble chaotically. Like the toy top, it is this slower speed and rotation of the subatomic particles that also creates the attribute of seeming to be more solid and visible to the naked human eye, because suddenly light waves can reflect off those unstable and slower subatomic particles of Physical Matter.

The slower internal speed is also what creates the slight distortions or wobbles in the patterns of internal vibration within the subatomic rotations occurring inside any Physical Matter particles, and the distortions cause the inherent chaotic nature. This creates the force of Chaos, which can only be measured with statistics and probability theories, that are only found within the classification of Physical Matter.

Ethereal Energy however, has subatomic particles that spin, vibrate and rotate at speeds that are much faster than those of Physical Matter, in fact at speeds greater than the speed of invisible visual light (C), so we can put them into the classifications of Ethereal Energy, which is the force of Order.

Their extremely fast internal speeds of the spin, vibration and rotation of all the subatomic particles of Ethereal Energy are much faster than that of Physical Matter and this is what prevents it from being easily seen with the human eye, because visual light is not able to be reflected off them and back to the naked human eye.

The faster internal speeds are what also creates smoother and more consistent patterns of internal vibration and rotation, which creates the force of Order, which can be seen within the multitude of mathematical formulas and precise equations that are only found within the classification of Ethereal Energy.

To summarize, the slower internal speed or vibration of Physical Matter creates slight distortions or Chaos which causes visibility and solidity but also creates defects and the constant change that is inherent within any Physical Matter. On the other hand, the much faster internal speed or vibration of Ethereal Energy creates smooth consistent vibration or Order which is not reflected and therefore invisible to the human eye, but which also prevent its solidity and makes change to those particles extremely difficult, at times almost impossible.

The nature of Chaos is that the Physical Matter is always eventually trying to reach entropy, so it degenerates, dies and falls apart, again due to its slower internal speed, structure and slight distortion in the internal spin of its subatomic particles. For example, if you take an automobile and put it in an empty field of grass and then leave it there for a thousand years, then eventually the **Physical Matter** that is the automobile, will rust and break down and degenerate, in much the same way that the physical bodies of humans will also eventually grow old, degenerate and die. Nothing in the physical world ever stays the same and most of the time the changes involved are rapid, unexpected and again chaotic.

The fact that the **Physical Realm** is truly chaotic is also why it also creates fears and worries within the human race, especially about physical death which can happen to any of us at any time without planning or any foresight. **Chaos** however, also creates the desire for individualism and freedoms, liberties and rights. It is the actual reason that decisions and **Free Will** exist in the **Physical Realm**. Physical creatures with intelligence can choose and make up their own mind about a great many things, even though some of the decisions will also be unexpected or chaotic in their own way and of their own choosing.

Although there are minor cases of **Ethereal Energy** changing, it can never degenerate or even be destroyed due to its internal steady structure and steady vibration that follow specific rules. Even the minor changes that can happen within the Ethereal Realm, such as electricity creating magnetism as it moves, must still follow very specific rules and regulations. Since all types of Ethereal Energy must follow the Order that has been established, there will be no sudden unexpected deviations that cause fear or worry. Order creates stability and uniformity but also restricts or even prevents freedoms, liberties and rights, as there is no room for individual decisions within the strict governance that is Order.

Within our own solar system, Order can be best seen with the example of the series of energy processes that occur on the surface of our Sun that brings forth invisible visual light, radiation and heat in an ordered and repeatable manner. Whereas Chaos can be best seen with the example of our Earth, simply since it became a broken planet in the near past, with tectonic plates that still shift and move as if they are still trying to settle and find equilibrium. In fact, if you remove the oceans and seas from the solid matter called Earth, then you would find that sometime in our very near past, a huge piece of a normally round planet has been broken and removed in several locations where the oceans of the world are present today. We are not living on a round planet made of rock like most other planets, but rather our planet is only disguised as round, due to the vast amount of water called oceans, that fill up all the broken areas, where land used to be.

Lastly, it is important to note that mathematics as an idea or process has always been a major part of the force called <u>Order</u> because all forms of <u>Order</u> are so stable and predictable that they can be shown with exact equations and formulas without worry of deviations or tolerances. For example, we already mentioned that the presence of moving electricity will create or convert to magnetism, but it will always be an amount and direction of magnetism that can be exactly calculated and demonstrated without error. There is no random variation present within the force of <u>Order</u>. However, we should note that some randomness will occur when measurements of <u>Order</u> or <u>Energy</u> are affected by <u>Chaos</u>, such as the measurement of electricity through a physical copper wire, because the measurement may vary but only due to the chaotic nature of the physical wire present, but it will never be due to the ordered value of the actual electricity.

New Unified Theory of Physics - Order overlaps Chaos

The second part of the new unified theory of physics continues out into Outer Space and the universe at a macro level.

POSTULATE 2 - Outer Space and the universe in general, may look invisible but is actually full of Order or Ethereal Energy in the many forms of electricity, radiation, magnetism, etc. mixed with physical invisible gases but also visible Chaos or Physical Matter, such as planets.

The Ethereal Energy within the universe is like a three-dimensional invisible pool or ocean filled with invisible sponge of energy that sorts, organizes and regulates the Chaos that is Physical Matter within it. This vast amount of free energy that exists as the Ethereal Energy system, in all the invisible space throughout the universe was originally called the Ether or the Prana in very ancient times and was known by the ancient people as the all-pervading vital energy of the universe.

The two separate forces, <u>Order</u> and <u>Chaos</u> are in constant conflict but as mentioned earlier, are also blended together and both directly impact each other under conflict and opposition but through attraction and seeking balance. All the <u>Physical Matter</u> that is the force of <u>Chaos</u> would be nothing but a big blob, but instead, it became sphere-like when it formed into planets and solar systems. The structure of physical matter within the universe became structured based on the invisible <u>Ethereal Energy</u> or the force of <u>Order</u> that surrounds it, much like the water of the ocean helps to structure the tides and the waves that carry the fish and aquatic creatures within it.

As described earlier, all subatomic particles vibrate, rotate, create torsion and vortexes. But in addition to that, <u>Ethereal Energy</u> doesn't just vibrate, those vibrations also modulate or in other words, its internal rhythm or frequency can combine with other frequencies and thus create slightly different pitches, frequencies, vibrations and rotations when combined. To use an example from earlier, as electricity moves throughout the universe at a certain frequency, rotation and spin, its electrical movement also creates or part of it changes into magnetism which moves at a different frequency, rotation and spin, then these two sub-forces of <u>Ethereal Energy</u> cushion or amplify each other, as they both move, which is modulation that occurs between them.

Remember that the universe is not an empty vacuum of nothingness, but rather all of Outer Space is filled with an invisible pool or ocean of moving and spinning <u>Ethereal Energy</u> that creates layers and pockets within the thinnest and thickest areas of the universe. The description used earlier of an ocean-sized sponge is kind of accurate in that there are pockets within this ocean of energy and within these energy pockets is where all the <u>Physical Matter</u> is located, simply present where they are. For example, planets, space dust, rocks or even suns are located within these pockets of the ocean-sized <u>Ethereal Energy</u> and as the energy moves, swirls and spins throughout the ocean of energy. So, it is this ocean of energy that is moving like water in a ocean, and that is what moves the planets around, just like a boat will be moved by the waves below it. The reasons that <u>Physical Matter</u> such as planets move throughout the universe in the way they do, at the speed they do, with the rotation and the pattern that they do, is because they follow the invisible pathways and pockets of <u>Ethereal Energy</u>. In fact, they must follow them and therefore are forced to follow these paths, because they are pushed along by the Ethereal Energy as it moves, spins and flows. This is unless of course, the physical object has somehow tapped into the energy source, that is the force of Order, and has somehow created its own path and speed throughout the ocean of energy, much like comets do.

Again, imagine an extremely large invisible sponge of energy, that is the size of the universe, but this sponge is actually made up of a free-flowing energy. Kind of like a massive three-dimensional river that is full of many small and large pockets of air, but the entire sponge is moving spinning and rotating. The small and large pockets are filled with Physical Matter, and these visible objects rotate, spin and move not because they have power or inertia but rather because the ocean-like energy is moving and rotating, which push around the planets, moons and suns in circles and ovals, by the invisible energy sponge that surrounds them. You can see how Chaos is formed, pushed, moved, spun and swirled throughout the universe, even when Chaos is created, within the same moving and spiraling invisible pool of Order, by viewing the same circular pattern of sea shells that are almost always the same pattern as the physical galaxies that form anywhere within the universe.

New Unified Theory of Physics - Matter does not pull as Gravity

When current scientists review the universe today, they usually try to explain the movements of the planets, stars and moons from the perspective of only the **Physical Universe** which forces them to create incorrect ideas and assumptions such as the idea that planets and moons rotate and spin, based on some sort of internal power that comes from the actual physical objects themselves. This incorrect perspective usually results in incorrect assumptions, for example the mistaken idea about the force of internal gravity within all matter that does not actually exist. Since scientists are only looking at the physical universe, they have developed the **Incorrect Theory of Gravity** that says that every piece of matter exerts a force upon all other pieces of matter. The larger the piece of matter, then the greater the force and this supposed force of gravity pulls at all other pieces of matter, moving them onto it or at least toward it.

If you think about this idea rationally however - if gravity was to actually exist, then it would require some sort of immense energy supply inside each piece of matter and that energy source would have to be immense, since it would need to operate every second, every minute, every hour and in fact forever. So, where is this energy source that allows gravity to exist? Well, scientists today say that they don't know, and they struggle to explain it, mainly because the force of gravity really doesn't exist at all. Even if there was an internal energy source for all matter, then how could it be sustained? The energy supply would be used up and must be replaced, so it makes no sense to assume that there is an energy source called gravity that allows planets and moons to rotate and revolve around the universe and it has its own power supply that never diminishes or gets used up?

It makes total sense however, if instead you think about the <u>Eternal Conflict</u>, which includes <u>Chaos</u> and <u>Order</u>. Think about the entire body of <u>Ethereal Energy</u> within the universe as a moving river or ocean of energy that pushes down upon all the <u>Physical Matter</u> that is trapped within it, within the small and large pockets inside the moving invisible energy sponge, then it is not gravity as an internal power that allows matter to pull at other matter, but it is rather the limitless and invisible external <u>Ethereal Energy</u> that is pushing down on <u>Physical Matter</u> and is actually doing the work. As the <u>Ethereal Energy</u> rotates, spins and moves then it also holds all the physical planets, suns and moons in place and as the invisible ocean of energy moves. <u>Physical Matter</u> does not pull upon all other matter, but rather the <u>Ethereal Energy</u> system pushes down upon all <u>Physical Matter</u>, holding it in place. Since balance must be achieved, the smaller the physical planet or moon, the smaller the required energy push that is required to hold it in place.

When you drop a ball, the ball starts to be pushed toward the planet by the <u>Ethereal Energy</u> pressure that is pushing down upon it. Much like atmospheric pressure, this <u>Ethereal Pressure</u> from the energy pushes and causes the object to start falling toward the planet, but also causes it to accelerate until it reaches the same downward speed as the constant <u>Ethereal Pressure</u> being exerted upon it. When we say that objects fall on Earth at the rate of 32 feet per second per second, we are saying that the <u>Ethereal Pressure</u> against the Earth is also 32 feet per second per second. It would be different on other physical planets or moons based on their size and mass. This idea of the <u>Ethereal Energy</u> pushing down on all <u>Physical Matter</u>, also explains why almost all universal matter is spherical in shape, because back in the very beginning when all physical matter was formless, the constant pressure from all directions against that matter would automatically form round balls, must like rain drops or soap bubbles form today, because of atmospheric pressure. There is also the idea that maybe the **Ethereal Pressure** from the Universe, which pushes down on each planet may also be the same as the atmospheric pressure that we measure daily, or at least is the cause of the atmospheric pressure from above.

Again, it may look like the Earth travels by its own power around the Sun, but in reality, the invisible **Ethereal Energy** is actually pushing the Earth around the Sun and also causing the Earth's to revolve around its axis and for the Sun to rotate around the Milky Way galaxy and also for our Solar System to rotate around the universe. The Earth, much like other rotating planets, spins around and through the **Ethereal Energy** electrical ocean and it is this movement of electricity against the surface of the Earth that literally create the magnetic currents and magnetic poles that exist around the Earth, it is not the Earth that creates its own magnetic field.

This is also the very reason why almost all of the solar systems that are known (including ours) have planets or planetary objects that follow a straight elliptical path or plane out and away from their sun or star. They all follow the same elliptical plane, much the same way that when you drop a pebble into a pool of water, the waves of energy and empty pockets within the waves ripple outward like the waves of water from the pebble in the same plane. The invisible energy sponge forces the physical planets and moons into empty pockets and then spins them outward in a similar path and plane away from their sun or star. That is why the invisible <u>Ethereal Energy</u> forces the planets to revolve on that same invisible energy path and plane and in the same rotational direction. Note however, that there are a few planets or moons that rotate differently, spin on a different axis or rotate on a different plane, plus there are comets and objects that may travel faster or slower with strange rotations and paths - but this is because the object has either tapped into the Ethereal Energy around it and it using this free energy for whatever reason or the object has been disrupted or altered by a different planetary object and so now that planet is stuck in a strange rotation, trapped within the spinning ocean of energy. Since around 90% of the planets, stars and other objects follow the flow pattern and straight outward plane of the Ethereal Energy around it, then any planet or moon that does not follow that path or rotation must have been disrupted at some point.

Once again as mentioned earlier, the invisible Ethereal Energy sponge or ocean may be three-dimensional but it also does not stay stationary, but rather like everything in the universe, it also moves, spins, vibrates and rotates, as each of the lines of energy flow in and out, down and up throughout the universe. The chaotic physical universe is being forced to follow a somewhat constant pattern and speed. It looks to the normal human eye as if the planets and stars move by themselves, but they are being forced to move, pushed as they rotate and forced to spin, as they are trapped within the invisible pockets that ripple outward.

This is also why we see far away galaxies as spinning clusters of stars and planets and spinning nebulas, because the invisible Ethereal Energy that fills the universe spins and rotates and causes these changes to happen. Since the Ethereal Energy is everywhere throughout the universe and alters Physical Matter through its attraction, it is this energy everywhere that also forms the famous formation called the <u>Golden Spiral</u>.

Ethereal Energy pushing onto the Planet in all directions and creating pressure.

PLANET

Planet trapped within an empty pocket of Ethereal Energy and moving and rotating with the wave.

STAR or SUN

Rests at the vortex or center of the moving and spinning Ethereal Energy wave.

Universe is not a vast emptiness, but rather full of invisible Ethereal Energy (ORDER) in the form of electricity, magnetism, radiation, etc. This Ethereal Energy moves, rotates, spins and has torsion much like a river flows.

Empty pockets within the Ethereal Energy fill with Physical Matter (CHAOS) and become spherical shapes due to the Ethereal Pressure being exerted onto the Physical Matter in all directions.

The Physical Matter is then moved throughout the universe by being pushed and forced in circular patterns based on the constant movement of the invisible Ethereal Energy.

A piece of Physical Matter, called a Planet, is trapped within an empty pocket that is found within the invisible Ethereal Energy that we think of today as the nothingness of Outer Space, but is actually full of massive amounts and different types of energy.

The invisible Ethereal Energy is the exact opposite of Physical Matter and since opposites attract - it will pushes against the Physical Matter, called a Planet, in all directions with the same Ethereal Pressure, forming a sphere and holding people and other objects to the Planet.

Since the piece of Physical Matter, called a Planet, is trapped within the invisible Ethereal Energy it will then be forced to move along the path of the flowing and spinning energy, much like a grain of sand moves downstream within the water of a moving river.

The planet's orbit will be in the same circular flow as the Ethereal Energy flows and in fact, all pieces of Physical Matter called Planets that are trapped within different empty pockets that will line up in a flat plane because of the flow of energy forces them to.

These various Planets line up but also emanate outward away from the center of the Ethereal Energy spin, in energy waves much like water waves that move outward, when a small pebble is dropped into a pool of water.

The movement, spin and torsion of the Ethereal Energy flows like a river and moves the Planet around within its flow and forms the orbit of the Planet. It also sometimes causes the Planet to spin on its axis, like a toy top.

When a Planet spins on its axis, like a toy top, then the Planet becomes an electromagnetic battery, with a north and south pole, because it is spinning within massive amounts of Ethereal Energy. When electricity moves, it creates magnetism.

The center point or vortex of the moving and spinning flow of Ethereal Energy usually results in the gathering of a tremendous amount of Ethereal Energy at that vortex, much like the center force of a whirlpool. That gathering of Ethereal Energy forms what we call stars or suns.

The energy that a star or sun uses to create heat and light comes from the actual flow of Ethereal Energy that condensing at the vortex of the spinning wave. The Ethereal Energy flows into the star of sun and allows it to burn creating heat and light.

The <u>Golden Spiral</u> is a logarithmic spiral that can be shown through geometry and mathematics to be the perfect golden ratio. All things in nature, whether the spiral form of a sea shell, the spiral form of deer antlers, the spiral form found in the leaves of trees and even the spiral arms of galaxies, all follow the same pattern, the same rotation and the same design. This is simply because all <u>Physical Matter</u> must follow the same pattern and design. This is one of the proofs that <u>Ethereal Pressure</u> is constantly pressing down onto all the different forms of <u>Physical Matter</u>, which instantly form spheres or gradually molds matter into the shape of the Golden Spiral.

New Unified Theory of Physics - Patterns and Cycles

Before leaving postulate two, it is important to also understand that much like the Golden Spiral, as the perfect ratio which is gradually formed into all matter by the <u>Ethereal Energy</u> pushing down onto it, there is also all the patterns and cycles that we see through our world and throughout the universe, which are also caused by the <u>Eternal Conflict</u>, and can be seen as proof of its existence.

Author Joseph P. Farrell, in his book <u>Babylon's Banksters</u>, discusses a similar idea to one put forth in this book, which is that any society that puts forth a <u>Chaos-based</u> type of culture, will also develop a <u>Chaos-based</u> system of money and finance but also at the same time that they will put forth <u>Chaos-based</u> types of science, physics and energy-systems. If you review the <u>Scale Chart</u> that is shown in the beginning of this book, you will see that I completely agree with Joseph Farrell but with one small difference and that is that I believe that the <u>Eternal Conflict</u> literally forces all aspects of civilization, society, culture, money, religion, construction, politics and even certain types of people within that civilization to move together as a complete unit. As civilizations rise and fall, they are molded by whatever primary force (either <u>Chaos</u> or <u>Order</u>) is dominant at that time. Whenever the force of <u>Chaos</u> rises, then the scale of <u>Order</u> becomes limited and off-balance and civilization changes to match the force of <u>Chaos</u> as it grows stronger. Whenever the force of <u>Order</u> rises, then the scale of <u>Chaos</u> becomes limited and off-balance, which then forces the civilization to change again to match the force of <u>Order</u> as it grows stronger.

Joseph P. Farrell, within his book, also discusses a relatively unknown department of the United States government that was created around the time of the Great Depression. The name of this agency is the <u>Foundation for the Study of Cycles</u>. Created under U.S. President Herbert Hoover to try and help understand the Great Depression and the U.S. Stock Market crash of C.E. 1929, the agency headed by Edward R. Dewey was not concerned with "the *whys and causes* of depressions, he decided to study *how* economic and business cycles occurred. He discovered that cycles, *waves* of behavior, existed in almost all aspects of economic and human social life, from the prices of pig iron to human emotions themselves". (Farrell, Joseph P. 2010)

Everything that the <u>Foundation for the Study of Cycles</u> discovered was published in a volume entitled <u>Cycles: The Science of Prediction</u> co-authored by both Edward R. Dewey and Edwin F. Dakin. They discovered that there were 3.5-year cycles, 9-year cycles, 18-year cycles and 54-year cycles that encompass literally everything: wholesale prices, wages, investments, product prices, commodity costs, manufacturing, industrial production, housing, marriages, religious values, etc. "What emerged from all this vast accumulation of data was that the growth curve of things – no matter what they were – was the same, and that there were discernible wave forms, cycles, of varying years' duration … these cycles were not only discernible but quantifiable because they were regular, and because they were regular, they were not only inevitable but also predictable … This is turn would seem to limit the scope of human action and its ability to influence such cycles". (Farrell, Joseph P. 2010)

I would add within Postulate 2 that these measurable, demonstrable, predictable cycles and patterns, that automatically occur, happen directly as a result of the Eternal Conflict and as Joseph Farrell stated, they are not based on human activity or even with human ability to choose. In addition to that, both Edward R. Dewey and Edwin F. Dakin in their book reflect on the question of what kind of hyper-dimensional physics could be operating behind the scenes and causing such inevitable and predictable cycles or trends to be forced upon humanity over and over again.

"In other words, not only are Dewey and Dakin suggesting that the sought-after deeper physics underlying their harmonic overlays and modulations of cycles might lie in a hyper-dimensional physics, but they are also suggesting that cause/effect thinking … might all be the immediate, though admittedly interrelated, *three-dimensional effects of a hyper-dimensional cause*". (Farrell, Joseph P. 2010)

To summarize, the Eternal Conflict and its only two primary forces are at work in the universe and cause Chaos-based societies and Order-based societies to rise and fall behind the scenes and when that happens, all aspects of those societies also change in harmony. With that being said, the third part of the new unified theory of physics can now move closer to a Planetary Level.

POSTULATE 3 – As mentioned earlier, gravity does not exist as defined today. Today we think of this mistaken idea called gravity as all matter having an internal force that pulls other matter towards it. This is not true, but rather what happens is that the invisible Ethereal Energy, throughout the universe, is attracted to matter and literally pushes down on matter in all directions, usually forming spherical shapes and holding objects onto planets through Ethereal Pressure.

We have already touched upon this subject earlier, but I have specifically included it again as Postulate 3.

This is because, the mistaken concept of gravity as a scientific explanation is one that Isaac Newton, Albert Einstein and even physicists today still have trouble describing let alone understanding. Also, as mentioned earlier, scientists know that if such a concept of gravity existed, it would require extremely vast amounts of energy to continue to be present in all matter, especially being operational forever in everything from planets to people.

So the physicists of the past and even those of today struggle to describe reality, because they do not understand that the <u>Physical Universe</u> moves and rotates based only on the <u>Ethereal Energy</u> systems that are invisible and exist only as three-dimensional grids, in and around the <u>Physical Universe</u> and in fact help to contain the <u>Physical Universe</u> that is caught within these invisible sponge oceans full of moving rippling pockets of energy.

To demonstrate further proof that gravity does not exist as presented by science, especially as a pulling force that emulated from all matter, then our moon should be slowly moving closer to the Earth, since the Earth has a larger mass and should have a larger pull on the moon.

However, this is not the case and in fact, the distance from the Moon to the Earth has been getting larger. The moon is moving away from the Earth and even stranger, the entire universe is also expanding outward and getting larger. So, the idea of gravity is kind of absurd, because all matter is not pulling on other matter, but instead away the spinning ocean of <u>Ethereal Energy</u> that makes up the universe and holds all the planets in place is gradually expanding as it spins outward.

Again, using the idea of the Earth and the Moon together and also if you think of the Earth as trapped within an empty pocket on the <u>Ethereal Energy</u> grid and the Ethereal Energy is like a huge river, constantly moving, spinning and rotating throughout the universe, then the Earth is moving and spinning because it is being carried along within this river, in the empty pocket of the invisible energy grid. Once you can visualize that, then you should be able to see that the Moon also travels where it is located, not because it is dependent on a pull from Earth, but rather it is just another piece of matter trapped within a smaller pocket, within the Ethereal Energy grid that spins like a river, around the Earth and around our solar system.

In fact, gravity as described by modern science does not exist and <u>there is no direct pull relationship between any two pieces of matter</u>, so the distance between the Moon and Earth has no direct relationship and that distance could get larger or smaller based only on the invisible flow of energy around them and of course the pockets that they float within. There is <u>only one push relationship</u> and it is between <u>Chaos</u> and <u>Order</u>.

When a person moves from the environment of Earth where they have weight and mass and then travel into Outer Space, they become weightless or begin to float. The actual reality however, is that the person stills has mass and weight, but simply the environment has changed around that person. On Earth, almost all the <u>Ethereal Energy</u> is pushing directly down onto the Earth as it is attracted to a huge piece of Physical Matter called a planet. However, once a person moves into Outer Space, the <u>Ethereal Energy</u> that is still pushing on the person changes but is no longer pushes straight down, but rather it is now pushing on the person from all directions, much like a small grain of sand in a moving river. This causes the person to start to levitate or float and then the person will start to move within the <u>Ethereal Energy</u> that is also moving, spinning and rotating throughout the universe.

This idea also explains why electricity as a part of the force of Ethereal Energy wants and needs to ground itself out directly to the Earth or to Physical Matter. Today we use circuit boards and electrical circuits as a way to trap and use electricity for our own use, even though electricity in general, only likes to control itself and does not want physical matter controlling it. However, for electricity to be used, we must give it a Pathway to Ground, to the Physical Matter called Earth, as that is where its attraction is anyways. Electricity will follow a path that we set up, but only if the result allows it to be grounded to Physical Matter, which is where the force of all the Ethereal Energy wants to go and is attracted to, pushing down on matter and as a side effect, ends up holding physical matter in place and creating its shape and movement.

This idea, finally explains the unusual scientific question of why a rocket or a spaceship that rotates while it rises up, will actually launch farther and higher than an identical rocket that uses the same amount of fuel during takeoff, but does not rotate as it rises.

Scientists have never been able to explain away this phenomenon, especially within their incorrect concept of gravity pulling downward on the rocket. However, with the truer understanding that the rocket or spaceship is rotating up and through the invisible **Ethereal Energy** that is pushing down on the spaceship, while the spaceship itself is also moving and rotating in the same spinning direction as the **Ethereal Energy** is also rotating. With that idea in mind, it then makes complete sense that the rotation would create a smoother path with less resistance for travel upward, so it therefore requires less energy and allows the rocket or spaceship to travel farther with the same amount of fuel.

Today, scientists use ideas of black holes to try and maintain the false idea that gravity exists, although the question of how a black hole or an area of super-dense gravity would work cannot be explained. If black holes did exist within the universe, then they are simply just remaining portions of the universe that are not in balance, where only **Ethereal Energy** exists but with almost zero matter to balance it. These are like the areas at the edges of the known universe that are still in-flux and out of balance and where you can see electrical storms, plasma discharges and vast nebulas.

In the beginning of time, those same electrical storms and plasma discharges also existed in the center of the universe and even here on Earth. That was because as you move closer to the beginning of time, then you can see that the entire universe was not in balance at all and in fact, everything was in flux. That is why in the earliest days of creation, we see cave drawings by primal humans that show electrical bursts in the sky and electrical or nuclear explosions all around them. These cave drawings show the two forces, <u>Order</u> and <u>Chaos</u> that during the beginning of time were completely unbalanced and destructive.

Since that time, most of the known universe, specifically the Milky Way galaxy, has found a sort of balance, where material and energy are roughly equal and somewhat in balance. However, there are still many areas around the universe that are still in flux. If one of these areas had too much <u>Ethereal Energy</u> and not enough <u>Physical Matter</u>, then a sort of black hole area is present. For more information on the early cave drawings that showed the electrical bursts in the sky and the presence of electricity around the Earth, please read

"Thunderbolts of the Gods" by David Talbott and Wallace Thornhill. (Talbott, David; Thornhill, Wallace 2005)

Physicists today have said that they have found gravitational waves that were measured coming out of black holes, but this is actually not what they are seeing and measuring, because those supposed black hole areas are simply a part of the universe out-of-balance that has more Ethereal Energy than matter and so what the scientists were actually measuring was the rotating and spinning waves of Ethereal Energy rippling outward from these areas, attracted to Physical Matter and moving in waves spinning outward.

New Unified Theory of Physics - Equation of their relationship

This new unified theory of physics presents a possible equation to show the relationship between Order and Chaos, or between energy and matter.

One of the most famous equations to date, from modern day physics was imagined by Albert Einstein and was written as $E = MC^2$ which is the equation that tries to explain how Ethereal Energy (E) relates directly to

Physical Matter (M). This equation was one of the earliest attempts to suggest a direct relationship between Chaos and Order by showing that Physical Matter can directly change to become pure Ethereal Energy. (Einstein, Albert, C.E. 1905)

The equation by Albert Einstein says that this can happen but only if the Physical Matter can be accelerated up to an extremely fast speed, which according to the equation is the speed of invisible visual light (C) multiplied by the speed of invisible visual light (C). This concept that Physical Matter or Chaos can become Ethereal Energy or Order helped to bring about the nuclear age that we live in today, by theorizing that physical atoms of matter can actually be turned into pure nuclear energy (which also demonstrated that sub-atomic particles of either physical matter or energy are the same particles but with just different vibration or rotating qualities). With all that being said, it is also very important to reflect upon the opposite viewpoint of this theory, because the equation (read backwards) also states that Energy or Order can become Physical Matter or Chaos by somehow slowing down. This change from energy to matter or visa-versa can and does happen also, based on The Eternal Conflict principles, except that I would take the stance that conversion from <u>Physical Matter</u> to Energy is much easier then the opposite conversion from <u>Ethereal Energy</u> to <u>Physical Matter</u>. This is because <u>Order</u> does not change easily or at all (its internal subatomic vibration is fast and stable), while <u>Chaos</u> changes constantly (its internal subatomic vibration is slow and unstable).

The other aspect of the famous equation $E = MC^2$ is that the relationship between <u>Physical Matter</u> called <u>Chaos</u> (M) and <u>Ethereal Energy</u> called <u>Order</u> (E) is completely dependent upon speed (C). Although most modern scientists and physicists today like to think in terms of external speed or external movement, I would like to suggest otherwise. If we minimize this idea down by thinking in terms of internal speed (or vibration) instead of external speed, then it may be easier to understand the paramount relationship between the force of <u>Chaos</u> and the force of <u>Order</u>. So, let us look internally.

I would propose that the speed that separates and defines the two known forces in the universe is its internal speed, spin, rotation and torsion of the subatomic particles within each force. Again, I propose that there is a limited set of subatomic particles and there is very little difference between the overall structure within the visible force of matter or the invisible force of energy, as they both formed with those same subatomic particles. The main difference being that each has a much different (faster or slower) internal subatomic speed, spin, rotation, torsion and vibration. The force of <u>Chaos</u> or <u>Physical Matter</u> has subatomic particles that move less than the speed of visual light and the force of <u>Order</u> or <u>Ethereal Energy</u> has subatomic particles that move greater than the speed of visual light.

Again, the Human eye cannot see <u>Ethereal Energy</u>, because visual light cannot reflect off the subatomic particles of <u>Ethereal Energy</u>, because they move and spin faster than visual light, so they exist right in front of you but are invisible to the naked eye. The subatomic particles of <u>Physical Matter</u> however, move and spin slower than visual light, so visual light can bounce off those particles and return to the human eye, hence physical objects are seen by the human eye.

In much the same way, the force of <u>Order</u> or <u>Ethereal Energy</u> does not have any mass or weight as it is moving superfast and moving in all directions usually against or toward the matter that it is attracted to. The force of <u>Chaos</u> or <u>Physical Matter</u> has subatomic particles that move and spin much slower and hence it has mass and weight.

One problem that Albert Einstein had with his formula is that he originally stated that the speed of anything can be relative and not absolute, such as the speed of a motorcycle at 50 MPH would seem to be 100 MPH to another motorcycle passing in from the other direction and while also driving 50 MPH. Albert Einstein, then later decided that only the speed of visual light would be constant. I agree with the idea that position and direction of the observer does change the perception of the speed that is being observed, but it still does not ever change the actual speed. Perception or observation may be relative, but the actual speed never is, because trying to combine observation with reality can sometimes be a problem and is the very reason that trying to study human behavior also becomes almost impossible. The actual speed of everything is real and not relative. It does not however, change the idea that the observation and the person doing the observing may perceive it differently.

POSTULATE 4 - I would also contend that **Physical Matter** or **Chaos** can become **Ethereal Energy** or **Order** if its subatomic particles can accelerate, speak up and approach the limit that is the speed of visual light. In fact, the quicker the acceleration happens, then the faster and more dramatic the change that occurs. This change can be seen in atomic or nuclear processes and is demonstrated by the following equation:

$$\lim (V \to C) = [(M \times (+\Delta V / T)) = E]$$

Where V is the internal velocity, spin, rotation and torsion of the subatomic particles, C is the speed of visual light, E is **Ethereal Energy**, M is **Physical Matter** and T is Time. So, the equation is described as the following: As the internal velocity, spin, rotation and torsion of the subatomic particles (V) approach the limit of C which is the speed of visual light, then the equation becomes a reality - **Physical Matter** (M) accelerates and becomes **Ethereal Energy** (E). The visible becomes invisible, the imperfect **Chaos** than becomes **Ordered** perfection and the piece of matter becomes a tremendous amount of energy.

As I mentioned earlier, it is much easier for <u>Physical Matter</u> to become <u>Ethereal Energy</u> because <u>Chaos</u> is normally in a constant state of change or decay anyways. Any form of <u>Radiated Matter</u>, such as Uranium that is naturally radioactive, would also be much easier to change into <u>Ethereal Energy</u> as it is already semi-blended with energy. Atomic reactions are one such example of <u>Matter that becomes Energy</u> and note that within atomic reactions, it is the *acceleration* of the internal vibrations of the radiated physical matter that defines its change to pure energy.

As mentioned earlier, it is very difficult for <u>Ethereal Energy</u> to convert itself to <u>Physical Matter</u>, mainly because it is very stable and does not like to change at all. But there is also the problem with trying to change (slow down) the perfect spin within the subatomic particles of <u>Ethereal Energy</u> by immediately decelerating or reducing its velocity, spin, rotation and torsion. The deceleration must be caused by a catalyst of some type and must be immediate, as you are taking perfection and causing it to wobble, corrode and spin erratically, which is taking the invisible perfection within <u>Ethereal Energy</u> and converting it to visible imperfect <u>Physical Matter</u>.

It is interesting that physicists today use electron microscopes and have seen instances where extremely small subatomic particles have literally appeared out of nowhere or disappeared to somewhere unknown place. They are described by the scientists are literally winking themselves in and out of existence. These instances have led some scientists to question or theorize about the validity of multiple dimensions. I would contend that these examples are simply the same effect that humans see with visible matter or invisible energy. As mentioned multiple times, subatomic particles move, spin and have torsion, so they can also independently be caused to accelerate and decelerate as they move. Particles moving faster than the speed of visual light disappear to the naked eye, since visual human sight requires visual light to reflect off the object to be able to see it. Energy systems such as electricity, radiation or magnetism are real but seem invisible, because of the speed of their internal subatomic particles. So, if one subatomic particle that can be seen with an electron microscope suddenly disappears, maybe the internal aspects within a particle simply accelerated to faster than the speed of the photons used by the microscope, or if a subatomic particle suddenly appears, then maybe it just simply decelerated to a speed that the photons could suddenly find it again and reflect it. If that was the case, then there would be no need for the complex idea of other multiple dimensions and the simpler explanation of change in speed (within the Eternal Conflict theory) would be sufficient.

Chaos will experience change but only change that is measured by the general concept of time, while Order almost never changes and therefore has no measurement called time. The concept of time, as human beings experiences it, is nothing more than the measurement of the change of something, such as the rotation of the physical Earth around itself divided into approximately 24 sections called hours and the rotation of the physical Earth around the Sun divided into approximately 365 sections called days.

Again, the Physical World and its measurements are never exact, hence the need to add or not add a Leap Year every four years, and the realization that time as a measurement of physical reality is actually never exact as everything within the force of Chaos has a wobble, including the planet Earth. Again, I would disagree with the scientific idea today that time is an actual dimension or just the curvature or a wave in space. Time is not a dimension, but rather just an abstract measurement device - much like a ruler or yard-stick, hence the idea or concept of time travel is simply just fiction.

It is important to note however, that there are two seemingly visual exceptions within the invisible Ethereal Energy classification. The first exception is visual Fire, which is created from the direct destruction of some specific types of Physical Matter such as wood or by a specific type of Ethereal Energy such as electricity.

The second exception is visual Light colors, which are created from the direct destruction of specific types of Ethereal Energy such as invisible visual light that is broken down into colors by a specific type of Physical Matter such as a crystal prism or drops of water.

These two items are seemingly on the borderline between the two major forces within the universe and that is why they react as such, seemingly in both areas of Chaos and Order at the same time, although they are not. It may be suggested that both fire and the visual colors seen within visual light may have a similar internal vibration or speed as their subatomic particles spin, vibrate and rotate or that their internal rotation is very close to the speed of visual light, so that they flicker back and forth as a force of Chaos and then a force of Order continually and seemingly on the edge of both forces.

New opportunity can be found within the new physics.

Having stated my ideas regarding the Eternal Conflict and some of the new concepts surrounding it, I also understand the controversy that comes with it. However, the largest revelations and breakthroughs in any branch of science have always come from without and not from within that group. The person that invented the copy-machine, a new kind of photography, was rejected by their fellow photographic networks. The person that brought the digital watch to the world was rejected outright by their fellow watch makers.

So, this new physics, which is the basic understanding that the world has only two fundamental forces, provides great opportunity to those willing to exploit it. If you understand that the invisible world around you, is not empty but instead filled with massive amounts of energy and energy systems, then you can also begin to understand that these energy systems can be tapped into and their energy used by anybody that wants to exploit it. The force of <u>Order</u> does not want to continually build up charges in the Earth's ionosphere that must be later be discharged as electrical storms and electrical lightning, but rather the force of <u>Order</u> is there for us to use, free of charge and pollution free, to the person that discovers the method to freeing it.

Nikolas Tesla, with his Wardenclyffe Tower in New York or Benjamin Franklin with his key and his kite, both realized these ideas and tried to tap into the source of all energy. They were unsuccessful for many reasons, but I do believe that someday, somebody will do this.

In the next chapter, we shall discuss the <u>Eternal Conflict</u> in more detail.

CHAPTER 3 - THE ETERNAL CONFLICT IN GENERAL

Chaos verses Order from the very beginning of creation.

The word <u>Chaos</u> comes from the ancient Greek word <u>Khaos</u> which meant the formless primordial soup, and which was all the <u>Physical Matter</u> that existed before the universe that we know of today.

The Holy Bible describes this <u>Physical Matter</u> that was pure <u>Chaos</u> in the first two sentences of the first chapter of the book of Genesis and which reads, "In the beginning, God created the Heavens and the Earth. The Earth was without form and void and darkness was over the face of the deep waters". (The Holy Bible, King James Version, Genesis).

This description is very interesting, because it specifically states that in the absolute beginning, the Earth was without form or formless, which basically means that it was not in the form of the sphere, as we know it today. Also, the Earth was void or had the appearance of nothingness. If these sentences were to be put into context then the word <u>Earth</u> as translated from the original text, cannot be describing the current planet that we live on. It makes much more sense if the word <u>Earth</u> was just a mistranslation that in reality described all of the <u>Physical Matter</u> that was the first primordial soup at the beginning of time. It may be better stated in a more correct translation as reading, "In the absolute beginning, <u>Chaos</u> or all <u>Physical Matter</u> was without form and void".

It is also interesting to note that throughout ancient history to even today, <u>Chaos</u> or <u>Physical Matter</u> has always been associated directly with the deep waters or water in general as the Biblical verse quoted above openly points out. The main reason for this association may be that most of the Earth is covered by water, which is part of <u>Physical Matter</u> and it is also through the water breaking and pouring forth that children are born, which is the process that new physical life is created from.

Within some modern dictionaries today, the word <u>Chaos</u> can be described as a failure to behave predictably or to simply behave differently than was expected. Within this definition includes the concept of free-will and freedoms, liberties and rights of individuals, which is the ability to choose a different path instead of following rules or regulations. The ability to behave differently than others or simply to do as you please, has always been a part of the Chaos which is the Physical Realm, and which is directly in conflict with Order or the Spiritual Realm.

On the other hand, the word Order can be defined as simply having a condition of logical arrangement where disruption is not allowed by an established authority. It is also defined as an authoritative indication to be obeyed such as a command or direction. There are also many other definitions but they all have the same things in common, which is that they always include stability, organizational authority and that of creating or following a set of rules, instructions or commands.

The Holy Bible in the same verse above also includes Order or the Ethereal Energy of the universe through the word Heavens. As we already discussed above, I believe that the word Earth was not the planet Earth but rather the formless and void Chaos or all the Physical Matter which was the primordial soup that started everything. The word Heavens would then simply describe all the Ethereal Energy that is Order. This sentence would be better read if it said, "In the absolute beginning, God created <u>Order</u> described with the plural word Heavens and <u>Chaos</u> described with the word that describes the known physical world which was simply Earth." Hence in the absolute beginning, the <u>Eternal Conflict</u> started as soon as both sides of our dualistic universe were created.

It is interesting to note that throughout ancient history to even today, <u>Order</u> or <u>Ethereal Energy</u> has always been associated directly with the wind, air or sky which is mentioned in the third sentence of that same chapter of Genesis that reads, "And the <u>Spirit</u> of God blew over the face of the waters", which alludes to the little bit of <u>Order</u> which was added to and blended into <u>Chaos</u> and which created defined shapes such as the round sphere of planets and suns.

One of the reasons for this association of <u>Order</u> with the wind, air or sky could be that most of the time that people actually see <u>Order</u> is usually during thunder storms or lightning storms that occur in the sky and in the air. It is also assumed that this lightning coming down from the sky also helped to create the first fires. Add to this the primitive understanding that our Sun brings forth life through its repetitive <u>Order</u> of bringing sunlight each day, which in turn, also led to the many stories where the spirits or Gods must be up in the sky or up in the Heavens.

As you read further, you will see that the ancient idea of <u>Chaos</u> being directly linked with water and that of <u>Order</u> being directly linked with air has always been a part of history, mainly because these associations not only started in the absolute beginning, but they have also continued throughout time, to this very day.

So, in the absolute beginning, the <u>Physical Realm</u> made of the force of <u>Chaos</u> was overlapped and blended with the force of <u>Order</u> distilled around and within it and that created all the ordered movement and defined shapes in the universe. Overall the randomness and vagueness of <u>Chaos</u> still reign in the <u>Physical Realm</u>. But at the same time, the <u>Ethereal Realm</u> made from the force of <u>Order</u> was blended with <u>Chaos</u>, which created minor randomness and inspiration.

<u>The Eternal Conflict and the misunderstanding of Good vs. Evil.</u>

The force called Chaos as one side of the Eternal Conflict loves both change and freedom, while it does not like rules or regulations to govern it. Chaos from a social standpoint is individualistic with many freedoms, liberties and rights. This means that its opposite force called Order wants stability, perfection and consistency. Order from a social standpoint will always develop people who want to control others or eliminate freedoms, liberties and rights for what they would call the greater good and for their desired social utopia.

There are many within the religious community today who describe the Eternal Conflict in turns of the struggle between the angels in the Spiritual Heavens and the humans on the Physical Earth. They explain this overall concept within that allegory by stating that since Earth is a part of the Physical Realm, then it makes sense that humans are given Free Will, which are absolute freedoms, liberties and rights given as a gift by God, to decide whether or not to follow God. They will also point out that humans are also given forgiveness, if they don't follow God but then later repent, because they have been knowingly placed within the realm of Chaos where mistakes are supposed to happen and are to be expected.

They continue with the idea that since the Heavens are a part of the Ethereal Realm, then it also makes sense that angels, as creatures of the Heavens, are not given free-will and are also not allowed forgiveness, because they have been placed within the realm of Order and must follow all of the rules without question, especially the regulations, rules and absolute laws of God.

Once again, the forces of Chaos and Order are opposite halves of the same duality called reality. They are mixed together while they oppose each other and are in perpetual conflict, even though they are both attracted to each other with a need for balance. Most of the religious organizations that have ever been in place also speak of the Eternal Conflict, but they incorrectly use the terms Good against Evil instead of the correct terms Chaos against Order. To me this seems to have been done deliberately but may be just misdirection since the terms Good and Evil are very generic term to prevent free thinking. They also know that the eternal truth about Chaos against Order is not generic but can be extended into everything we know and also understand which tends to keep the universal truth of the Eternal Conflict a secret. Again, the terms Good and Evil as used today by various religions as if they were absolute terms, those two words are just relative localized concepts with no universal understanding at all.

Remember that wherever there are groups of people, there also seems to always be both written and unwritten rules about behavior and conduct. These rules may be unwritten such as how you should greet one another or what kinds of words are acceptable for general usage. These rules may also be written down as in laws, statutes or rules that invite arrest or imprisonment. People that follow the unwritten or written rules are considered <u>Good</u> by that society and people that don't follow the rules are considered <u>Evil</u> by that society. Understand that what may be considered <u>Good</u> in one country or society may be considered <u>Evil</u> in another country or society, while other actions may be viewed as <u>Evil</u> in one group or religion may actually be rewarded as a <u>Good</u> thing within a different group or religion.

It is critical to your true understanding of the <u>Eternal Conflict</u> within our dualistic universe, that the concepts of <u>Good</u> and <u>Evil</u> are never absolute ideas but instead they are subjective concepts defined differently based on location and community or religious systems.

This misdirection of <u>Good</u> against <u>Evil</u> also prevents the side-by-side appreciation and true understanding of the many different ancient writings or religious texts which discuss the truth regarding the <u>Eternal Conflict</u>.

The spiral action of everything creates cycles and patterns

Many of the equations and formulas that are found within science and technology have been discovered through observation and study. This includes some of the <u>Ethereal Energy</u> that we are aware of, such as electricity, radiation and magnetism, but it is important to also note that there are probably many other types of invisible <u>Ethereal Energy</u> that we are not aware of yet. For example, many people believe in ghosts, demons or spirits and even lay lines, energy patterns or holy or haunted locations on Earth that have more or less <u>Ethereal Energy</u>. All these items seem strange or even supernatural, but none of them can be dismissed out of hand due to the lack of information that we have on all of the types of invisible <u>Ethereal Energy</u>. We just don't know if we have discovered all of them yet, especially since all the different types of <u>Ethereal Energy</u> entities are invisible to the naked human eye.

There is a theory in physics that I do agree with and that is called the Law of Conservation of Energy. This law states that the total energy of any system is conserved over time, which means that energy cannot be created or destroyed but rather only changed. This law prevents a perpetual motion machine from ever being invented, which I also happen to agree with. Almost all scientists also believe this law and if this law is true, then who was are as humans can never be destroyed. That is because the electricity that is found within every human brain, traveling up and down the human spinal column and throughout the human body must go somewhere after the physical body dies, because the energy that is our thoughts and ideas cannot be destroyed but rather change to something else and move on. So, in my opinion there must be an afterlife for our thoughts, ideas, spirits and souls, even though we currently have no data yet to confirm exactly where it is.

One thing that we do know about Ethereal Energy is that it never stops moving. It also never moves in a straight line, but rather vibrates, spins, rotates and creates torsion while moving in a spiral action, much like a spring. In fact, you can see the spirals and swirls of the invisible Ethereal Energy surrounding us and is the actual causation of many of the visible things that we see everywhere within the physical world. As the swirling energy pushes against the physical world, we see whirlpools of water, spirals in seashells, spiral flowers, twisted horned antlers of deer and even the twisted double helix of the DNA of all matter. When you look throughout the universe, you also be spiral swirling galaxies and billions of rotating and spinning solar systems. Literally everything observed within both the force of <u>Order</u> and within the force of <u>Chaos</u> spins, rotates, vibrates and creates torsion.

There are many different ideas regarding the movement of electricity based on their current or amperage. Some say that direct current (DC) electricity moves straight while alternating current (AC) electricity moves up and down, back and forth. I believe that both of these ideas are wrong.

My brother Richard Biers also speculated to me that direct current (DC) electricity moves in an extremely small tight spiral, like an extremely tiny spring, that can be viewed as being on a straight line when viewed on a two-dimensional oscilloscope. This spiral is so tiny and wound so tight that when it travels through a controlled conductor such as a wire, it will travel deep within the wire and is therefore almost harmless to human touch. This type of electricity however, creates lots of friction and resistance because it travels deep within the copper wire. (Biers, Richard Lee C.E. 2011).

On the other hand, alternating current (AC) electricity moves with a very wide loose spiral, like a large spring that is very wide and when viewed on a two-dimensional oscilloscope is viewed looking like a sign wave with just an up and down motion. This spiral is so large that when it travels through a controlled conductor, such as a wire, it travels by spinning around the surface of the wire and is therefore extremely dangerous to human touch. This type of electricity creates very little friction and resistance, because it just skims the surface of the copper wire. The fact that alternating current (AC) electricity has very little resistance is why Nikola Tesla was so successful in powering the world with his invention called alternating current (AC) electricity. (Biers, Richard Lee C.E. 2011).

Again, the spinning, rotating and moving aspect of both sides of the Eternal Conflict are what creates the patterns and cycles that we can see. Chaos or the

Physical World creates hard to analyze patterns and cycle with inherent randomness, while Order or the Ethereal Energy that exists creates very specific patterns and cycles that can be analyzed with exact formulas and equations.

The modulation and movement of all the invisible Ethereal Energy, throughout the universe, moves in a very organized and repeatable pattern. As described earlier, these patterns of energy flowing in and around the planets and stars can be charted and will reveal themselves to closely resemble sound waves. It is these repeatable patterns of energy flowing around us that also create the repeatable cycles that we experience in life, such as business cycles, economic cycles, sunspot cycles, financial cycles, agricultural cycles and even weather cycles. These patterns of energy are invisible, but they also flow in and around everything physical including us.

The Ethereal Energy Sponge – influences and free energy

Today, most people mock the subject of astrology, but it is important to understand that while seeming to discredit this subject, most of the very successful people in the world use astrology on a daily basis and they truly understand and use the universal cycles that are present on the planet and throughout the universe. However, they use the entire astrological system of planets in signs, mathematical degrees of influence each planet has based on position, positive and negative influences of each of the mathematical measurements, overall of houses on top of the positions, etc. The general population is only exposed to the terribly limited idea of only twelve zodiac signs in the newspapers, while others use the entire system of repeatable cycles and patterns in secret to help guide worldwide politics and especially international finance.

These tools are available because of the <u>Ethereal Energy</u> grid that is present throughout the universe, within the very atmosphere of our planet and within even the people on our planet. It guides the <u>Physical Matter</u> not only daily but even through some very important cycles that can be shown to be repeatable throughout decades, centuries and even millenniums. Real astrology from ancient Babylon can be used to discover these patterns and cycles and is the reason why newspapers will only print generic astrology columns that are meant to divert attention away from the repeatable cycles that can be used to gain and maintain power.

As mentioned earlier, the Ethereal Energy is everywhere and overlaps everything physical and it is this overlapping which is what helps to define the Physical Matter that is being pushed upon. This Ethereal Energy is invisible to the human eye and contains vast amounts of electricity and other sources of energy that are used to power suns and stars, instead of the incorrect idea that suns and stars fuel themselves. I believe that there is no fusion-able or fission-able energy source from within our sun, instead the sun uses the energy from outside of it, using the Ethereal Energy ocean to fuel itself. Therefore, when a burst of energy comes off the sun, it does not come off as a straight line but rather as an arc, two straight lines connected together at the top, which is very similar to electrical arcs. Also, the Ethereal Energy ocean moves against the Earth and is responsible for such things as the Aurora Borealis, which are the northern lights that show up over the Earth's north magnetic pole and the Aurora Australis, which are the southern lights that show up over the Earth's south magnetic pole.

For a minute, just imagine that the seemingly empty space around your body is filled with 70,000 volts of electricity, which is invisible powerful energy. You may not know this, but strangely if the voltage, in the air around you, stayed at exactly the same voltage (70,000 volts everywhere) than you would not even feel or sense the electricity, at least as long as the total voltage never changed. In truth, in order to hurt or damage a human body or in fact, hurt or damage anything in the <u>Physical Universe</u>, there must be a change in voltage or what is called a <u>DROP</u> in <u>Voltage Potential</u>.

The voltage potential of energy that surrounds us in the seemingly empty space is a constant static voltage and only would require a way to change the voltage slightly to create a usage stream of free energy. Think of it this way, certain garments and fabrics in extremely dry conditions tap into this <u>Static Electricity</u> that is all around us by slightly altering the constant state of potential. If you ever got zapped by even a small amount of static electricity when you for example, removed a wool sweater from a clothes dryer, then you should be able to imagine the vast amount of potential electricity that is in the air surrounding you.

The idea of free invisible energy surrounding you as <u>Ethereal Energy</u>, also supports the strange idea of levitation as a very real possibility. For example, if we could create extra torsion fields, especially using specific physical matter such as red mercury, that works for or against the static consistent Ethereal Energy voltage that surrounds us, then the torsion fields created by the specific matter may push away or eliminate the surrounding Ethereal Energy that usually pushes down on a object, thus creating an empty pocket or another sponge hole, that would then be filled with physical matter. The physical matter would seem like it was levitating up off the ground, when the Ethereal Energy would actually be what was pushed away and therefore could not push down on the physical matter for even a brief amount of time.

Both the magnetic field and the free energy on and around the Earth is an almost infinite amount of free power and it could be extracted and used by humanity without cost and would be pollution free. But using this free energy by the people of the world, cannot be allowed at least not while Order-People stay in control. The people that run the world today are mostly Order-People which work very hard to prevent the true understanding of the Eternal Conflict and the relationship between Chaos and <u>Order</u>. There will be more discussion in the next chapter about <u>Order-People</u> and <u>Chaos-People</u>.

CHAPTER 4 - THE ETERNAL CONFLICT INSIDE EACH PERSON

The Eternal Conflict within each individual.

Just as the Eternal Conflict is part of everything externally, this same duality exists within all humans internally. All humans have a material body made of Physical Chaos and an ethereal soul or spirit, composed of electricity, within that body made of Ethereal Order.

Humans also have a brain that is dual-hemispheric or has two different and separate pieces. There is a specific left-side of the human brain and a specific right-side of the human brain and both sides also contribute greatly to the dualistic viewpoint that we all have concerning our universe. In addition to having two sides to the human brain, the two different sides blend together but also differentiate and govern both aspects of the Eternal Conflict with the right-side of the human brain mostly governing Chaos and the left-side mostly governing Order.

The right-side of the human brain governs Chaos as it is responsible for creativity, emotional control, empathic relationships and inspiration. It is here that the need for individuality forms and the desire for freedoms, liberties and rights develop.

The left-side of the human brain governs <u>Order</u> and is responsible for rote memorization, logic, analysis, mathematics, language, technology and control. It is here that the need for order and social control forms and the desire to establish rules and regulations develop.

As the human brain grows and develops, there is always an internal conflict that gets created between both sides of the human brain and one dominant side will eventually surface, in which you will become either a more <u>Materialistic</u> individual following <u>Chaos</u> or a person that is a more <u>Ethereal</u> individual following <u>Order</u>.

The famous father of psychoanalysis, Doctor Sigmund Freud published his famous book "The Ego and the Id", where he used the term <u>Id</u> to describe the <u>chaotic right-side</u> and the term <u>Super-Ego</u> to describe the <u>ordered left-side</u> of the human brain. The <u>Id</u> side was the animalistic and chaotic side representing freedom, desire, risk, creativity and hope. It also governed subconscious thoughts. The <u>Super-Ego</u> side was the self-esteem and ordered side representing logical, plans, restrictions, organization and caution. (Freud, Sigmund, C.E. 1923)

It is interesting to note that there are many scientific studies that prove out some of Sigmund Freud's ideas and strangely, they have also proven that the <u>Chaotic</u> right-side of the brain controls the left-side of the Physical Human body, while the <u>Ordered</u> left-side of the brain controls the right-side of the Physical Human body. Sigmund Freud also described the human brain's relationship with itself as having a <u>Trinity</u> viewpoint, with the term human <u>Ego</u> to describe the relationship between the chaotic force called <u>Id</u> and the ordered force called <u>Super-Ego</u> within each brain.

One of the most important things to understand about people is that because the <u>Eternal Conflict</u> is present within the human brain, then both sides of the conflict are within each of us and that internal conflict also then becomes external also. If you slow down and pay attention to your thoughts and ideas as they form inside your brain, then you will notice that the two sides of your brain communicate and talk back and forth with each other. It may be difficult to realize this, especially if you have not developed this realization naturally. But if you really pay attention, then you will notice that both sides of the brain, the <u>Chaos</u> side and the <u>Order</u> side mostly argue back and forth as they are generally in conflict and continuously disagree with each other, even though they are working together to reach balance.

The <u>Chaos</u> side of the brain and the <u>Order</u> side of the brain argue back and forth, and it is only through this important internal dialog that you internally balance and can form opinions about your internal self or the outside world. This idea has been represented numerous times in history, most notably as having a small angel representing <u>Order</u> on one of your shoulders which usually suggests restraint or control of some type, while a small devil representing <u>Chaos</u> sits on the opposite shoulder and suggests mischief and randomness. Again, government and religious leaders will tell you incorrectly that this representation of the angel and devil is the conflict between <u>Good</u> and <u>Evil</u> but that again would be wrong, as once again the idea of <u>Good</u> or <u>Evil</u> are both subjective. The truth is that they represent the conflict between the <u>Id</u> and the <u>Super-Ego</u>, which is <u>Chaos</u> against <u>Order</u>. Both of which are absolutes.

In fact, most of mental illness that happen within the human brain can be directly or indirectly attributed to an interruption between the two sides of the human brain or when the two sides of the human brain stop communicating with each other. This causes a split down the middle of the brain and a separation of the <u>Eternal Conflict</u> which is critical to all human interactions both internally and externally. If you think about the names for most mental illness, such as bi-polar disorder or split-personality disorder, then you will see the absolute truth to this statement.

The internal discussion that should be going on between both sides of the brain has been credited for most of the great inventions of Nikola Tesla and the incredible mental abilities of Edgar Cayce, as they were unique in their ability to not only understand but to directly utilize this internal discussion ability between <u>Chaos</u> and <u>Order</u> to create balance.

Sadly, the American public government-run schools and public colleges of today do not even attempt to help children build up both sides of the human brain, but they rather focus through select educational techniques on developing only the left-side of the human brain, which crated citizens that are not good or capable of critical thinking and that do not have the correct balance between the two forces at work in the universe. Private elite schools and elite colleges and universities on the other hand do allow an education that will help to develop both sides of the brain.

Most of marketing and advertisers today want masses of people that are Materialistic consumers, so they influence government-run schools to focus public education on strengthening only the Order left-hand side of the human brain through rout memorization and mental testing and with very little focus on creativity and ingenuity which is the Chaos side of the brain. Like a reverse psychology experiment, this Ordered government-run educational experience creates off-balance situations within the human mind of the children at the public schools. Since Opposites Attract, the result is a child who desperately wants and desires Chaos experiences; such as physical enjoyment, instant gratification, sexual experiences, drugs and alcohol abuse.

Public government-run schools also focus on only one side, to quiet down or even prevent the internal conversation within the human brain from occurring. In fact, most people that were taught in public government-run schools are usually not even aware that the internal conversation within the human brain even exists or even happens. The result is masses of adults that are extremely Materialistic with numerous additions and habits.

Ethereal people against Materialistic people.

No matter what your educational experience was however, you need to understand that the Eternal Conflict exists internally within each of us and more importantly, that any individual, at any time, can add more Chaos to themselves and become more Materialistic, or they can add more Order to themselves and become more Ethereal or what can be called Spiritual. Note: The word Spiritual should not be confused with the word Religious, even though they are sometimes used together. The word Spiritual in this book is deemed to be a person that is Order-Based.

Adding more Chaos and becoming more Materialistic or adding more Order and becoming more Ethereal or Spiritual is the easiest when you are a brand-new baby that exists without established mental paradigms. This is mainly because during your formation years, you are still developing patterns within your brain to help you process information. The older we get as humans, the more difficult it becomes to try and change the type of person that we are or have become.

Individuals that become <u>Materialistic</u> by adding <u>Chaos</u> tend to only care about physical matter or the things within the <u>Physical Realm</u>, which is mostly visible and external from themselves. They usually care deeply about objects and physical things for themselves while also demanding freedom for themselves and sometimes others. They are also usually very emotional and tend to make rash decisions even though they incorrectly believe that they make logical decisions, which they don't. They also form their self-esteem and worth based on external things, such as sports teams they follow, types of automobiles that they drive or types of beverages that they drink. Their individual identity is all about objects and external things and it becomes difficult to try and change their mind about the brand loyalty that they have picked.

On the other hand, individuals that become <u>Ethereal</u> by adding <u>Order</u> have very little need for objects or physical things and may even give away their physical objects and even quit their physical job, unless those physical things help them with control. They tend to only care about rules, regulations and control while demanding strict conformity to their ideas. They believe that they are the only ones that use logic and reason and therefore seek out and sometimes even demand to oversee other people. They are also very internal and usually form a deep internal value system that may be positive or negative toward others. Most people assume that <u>Ethereal</u> or <u>Spiritual</u> people have high morals or ethics, and this may not be the case at all. They form their own internal rules and follow those rules, at least if they personally benefit. Their individual identity is based on internal items, such as their knowledge, their idea or their belief systems.

Once again, it is important to understand that the word <u>Spiritual</u> does not mean <u>Religious</u>. I mention this point again, as those two words today are confused on purpose by government and religious leaders. You can be very <u>Materialistically Religious</u> with external physical temples, expensive statues and lots of physical objects used for rituals and adoration, or you can be very <u>Spiritually Religious</u> with little or no physical objects or buildings worshipping outside or internally alone with silent prayer, much like a monk living high in the mountains. This subject of the **Eternal Conflict** within religions will be expounded upon later.

Generally, **Materialistic** people tend to be more chaotic and procrastinate more. They also tend to be more available, volunteer more, tend to break some rules such as speeding while driving, are more patient with others and are more open-minded. They can focus their mind onto one subject, but it will usually be physical in nature, such as machines, objects, sports or science. They tend to be workers that want to physically get things done and since opposites attract, they tend to be attracted to **Ethereal Leaders** that they will have always have conflicts with.

Ethereal or **Spiritual** people with more **Order** tend to be less available, view their time as more valuable, volunteer less, hardly ever give to charity, follow all the rules if they apply to them, are impatient with others and are closed-minded. They can focus their mind onto one subject, but the subject will usually be spiritual in nature, such as art, writing, drama, philosophy, mathematics, language or music. They tend to be leaders that can see the big picture and are usually attracted to Materialistic followers who they will always have conflicts with.

It is also important to note that there are many different shades of humans. For example, some people that are almost all Materialistic may be made up of 90% Chaos and 10% Order, while other people that are almost completely Ethereal may be made up of 10% Chaos and 90% Order. Of course, there is almost an infinite variety in-between those numbers and you can even add to that the fact that people are changing each day.

However, only rarely is there ever a human being that comes along that has a total brain that is exactly balanced 50% Chaos and 50% Order and then stays that way. These individuals, when they do appear, tend to be the prophets and holy people that can teach and preach actual reality and absolute truth so that everybody, including Chaos-People and Order-People can understand, because they are the only ones that truly comprehend both sides of the Eternal Conflict. It is also why most of those prophets and holy people that speak the truth tend to get killed quickly and violently because they show both sides of every issue and so nothing can be left hidden, which tends to make one or both sides angry.

The Eternal Conflict creates habits and addictions in Humans.

The human body has numerous receivers of information, such as ears to see, ears to hear, tongue to taste, nose to smell plus the millions of nerves for touch response. All these receivers together create a situation where our human brain always receives vastly more information, on a second-by-second basis, than can ever be processed completely.

To cope with this extreme amount of information that is being receiving constantly, the human brain develops broad patterns called paradigms that relate to different objects or situations. These paradigms are an electrical overlay on top of the physical brain, again Order and Chaos blended together. For example, a paradigm or pattern is developed to help the brain identify what a house is or what a car is. That way when all the input information reaches the human brain and it fits any of the normal patterns, then the human brain skips most of the incoming information regarding that object. It just recognizes the broad pattern and not the separate individual pieces of data.

Paradigms are very useful, and the human brain needs them as a filter. However, if a different kind of house is viewed that does not fit the normal pattern then the human brain tends to have trouble grasping the new object as a house. In some cases, the human brain will refuse to accept the object as a house and in extreme cases the human brain will literally be prevented from even seeing the object directly in front of them. Believe it or not, this happens quite a bit, especially when the information being received is not an object but rather as a new idea or a new concept. New ideas, new concepts or radical changes are almost always resisted at first and sometimes never accepted because of the way the human mind creates and uses paradigms. Materialistic <u>Chaos-People</u> tend to have more flexible paradigms. They also tend to be more open-minded and trusting while Ethereal <u>Order-People</u> tend to have more closed rigid minds that are more distrustful and hate change.

All of this together creates a situation whereas the human brain automatically develops habits and addictions. These habits and addictions stem from the multiple paradigms or patterns that get created to sort out the multiple input streams of information but also because the broad pattern creation is eternally in conflict. Habits are useful small patterns in life that help humans cope and they can be as simple as the how you take a shower, whether you wash your hair first or your feet first. They could be the route that you drive every day when going to work. All humans because of the human brain love and desire habits and create them all the time. According to Doctor Maxwell Maltz, it takes a human 21 days to develop a new habit and 21 days of doing something different to change a habit. However, this automatic desire and creation of simple habits based on the Eternal Conflict can easily lead to the next step, which are addictions. (Maltz, Maxwell M.D., F.I.C.S., C.E. 1960)

Addictions are extreme long-term habits that have moved from being a habit and changed into becomes a physical need or a spiritual need. These needs can become so extreme that the addicted individual will go out of their way to make sure they happen. If the addiction does not occur each day, then the individual can sometimes become physically ill, especially if the addiction is Chaos-driven. But sometimes the person can become spiritually withdrawn, especially if the addition is Order-driven. Chaos addictions are physical, such as alcoholism or drug abuse, whereas Order addictions are mental, such as Obsessive-Compulsive Disorder or even mental abuse.

Some habits and addictions can be simply the love of a sport or sports team, knowledge of trivia or playing guitar, but the truth is that all humans will eventually develop habits. However, once the habits become something that must happen each day, then they can be destructive as addictions. To minimize the daily habits that we have and to prevent them from becoming addictions, all humans must try to balance both sides of yourself, either physical or spiritual, every day. You must take care of your <u>Physical Body</u> as well as your <u>Ethereal Mind</u>. Again, the <u>Eternal Conflict</u> is always present within us and everywhere and the goal must be to find balance between both sides.

God-like objects and the worst addiction

The <u>Eternal Conflict</u> is everywhere and within everything. Since everything is a part of God, then both <u>Chaos</u> and <u>Order</u>, hence the <u>Eternal Conflict</u> is part of God. This statement is true, whether you are an atheist or are a religious individual, because people and objects can and do become God-like, if they have both <u>Chaos</u> and <u>Order</u> within them.

Let me explain by stating that any physical object is just an object unless <u>Order</u> or <u>Spirit</u> gets attached to it. Again, when I say <u>Spirit</u>, I am referring to an idea or a concept, a vision or an ideal. A piece of wood is just a piece of wood, however if somebody carves that physical piece of wood into a statue and then adds <u>Order</u> to it, by telling a story about it or attaching a mythology to it, then it has the potential to become God-like. If accepted by the masses of people, then it can become God-like but only because it has both <u>Chaos</u> and <u>Order</u>, which is <u>Physical</u> and <u>Ethereal</u> added to it, which are both sides of everything including God, if you happen to believe in one.

Objects and even people that have become God-like become objects of worship, which are some of the most forceful and vicious addictions. God-like objects or God-like people affect the mind and even alter perceptions, thoughts and even ideas. This makes them even more potent and forceful because another trait or habit within everyone usually also becomes the desire to alter the minds of other people, whether it is through movies, music, books, stories, computers, drugs or alcohol. God-like addictions that become accepted by a certain society can lead to death, destruction and even wars. The worse addiction of all time comes from a God-like object called Money.

Money will be discussed in more detail in a later chapter, but it is important to note that money is usually a physical thing made of Chaos, such as a metal coin or a piece of paper. But in addition to that, money is also Order, as the ultimate concept or idea. This concept of money causes people to dream of things to either purchase or places to visit that the perceived value of money allows this to happen. Some people will lose themselves both physically and mentally in the pursuit of more and more money. Some people will literally destroy others as well as themselves in the same pursuit.

Like the worship of God-like objects or God-like people are the large groups of people that come together for a specific worship of different concepts and that is mainly in the form of various religions. The next chapter will expound on these ideas and will discuss religion under the context of the Eternal Conflict even further.

Chapter 5 - TYPES OF RELIGION

The Dual nature of the various Religions

My personal belief is that there is a God but that everything is a part of God and that God is everything. Hence my opinion is that everything in the universe, including ourselves is a small piece of God. I also believe that in combination, both the force of Order and the force of Chaos, together make up what I think of as God. With all that in mind, I also believe that if God exists, then God would also be dual-natured and therefore the Eternal Conflict would also be a struggle for God, similar to the same struggle for humanity, both within and without.

With that in mind, I also believe that any religion will either be Order-based or Chaos-based depending on how they use their symbols, stories, mythologies and traditions to try and understand God but also to connect the Physical World with the Ethereal or Spiritual Heavens, which is the same as trying to balance both sides of the Eternal Conflict.

Whether or not a separate Deity exists called God or even whether or not you believe in God is not as important as the idea that the **Eternal Conflict** exists within all of humanity and creates two specific kinds of religious individuals and religious institutions. Although there are literally thousands of various religions within the world today, they can all be grouped into only two specific kinds. It is these two specific types of religions that have happened throughout ancient history even up to today and it is the conflict between these two specific types of religions that have created most of our historical wars.

The first kind of religious individual or institution is one that focuses on the **Chaos** or **Matter** side of God and of which believes in freedoms, rights and liberties. These **Materialistic Religions** see God as giving humanity a free-will or the choice to follow God or not. Their major belief is that those people that decide of their own free-will to follow God will be saved and those that decide not to follow God will be punished. The importance of a religion based on **Chaos** is that God is usually seen a friendly and forgiving mother or father figure. In **Chaos-based** or **Materialistic Religions**, God is there to help guide individuals throughout their lives with the understanding that people have a choice and most likely will make lots of mistakes, hence the forgiving side of God becomes critical within the worship of this version of God and the idea that everyone must keep an open-mind regarding everything, each other and even God.

Materialistic Religions are also based on **Chaos** or **Physical Matter**, so physical objects such as huge temples or churches with many statues and objects are very important. Some physical objects, such as crosses, candles, altars, thrones, dishes or chalices will not only be critical, but may also be covered in gold or silver to more adequately represent their materialistic value within the religion. In fact, everything of value will be external and even searching for God will be part of a **Chaos** religion and most likely result in an external search using other people's information to search for God, who will of course be located somewhere else like far up in the heavens or across the universe.

The second kind of religious individual or institution is one that focuses on the **Order** or the **Ethereal** or **Spiritual** side of God which desires strict rules, regulations and rules. These **Ethereal Religions** see God as only an authority figure that demands that humanity follow strict rules and regulations. In **Order-based** or **Ethereal Religions**, God is there to maintain obedience with the understanding that humanity does not have a right to disagree with God and God is seen only from a logical viewpoint with a very closed-mind, hence **Order-based** religions believe in repetition and habit as their main dogma as well as personal sacrifice and even pain.

Ethereal Religions are also based on **Order** or **Ethereal Energy**, so physical objects have little value and are sometimes not even a part of the religion. Knowledge and thoughts will become the most important thing, to which can be added ideas such as internal reflection, meditation and acceptance as being what is most valuable. The energy systems of the human body are within the brain and spine and so searching for God will only happen internally as the main attribute of **Ethereal Religions**.

Chaos-Based or **Materialistic Religions** usually have **Materialistic followers** that tend to be lazy. These are followers that do not want to be mentally stimulated or forced to study by themselves but would rather be told what to do hence **Materialistic Religions** rely on a **Faith-**

based system. Having faith in God requires no proof, zero facts and most importantly zero effort. I just have faith that what I am being told is true.

Again, since **Opposites Attract**, it is interesting to note that **Chaos-based** or **Materialistic Religions** tend to attract very **Order-based leaders** while the **Order-based** or **Ethereal Religions** tend to attract very **Chaos-based leaders**. This may seem strange and counterintuitive but is actually the norm for various reasons.

Order-based leaders are attracted to these **Materialistic Religions** because they desire followers that do not think for themselves but would rather just listen to the leader for advice and guidance. The **Order-based leaders** want to be in charge as the authority figure because they love control and order, so they love to lead **Materialistic Religions** because the followers are usually easier to control, and in some circumstances control themselves. This is like the idea that most political leaders today are **Order-based**, due to the Materialistic and consumer-based leanings of most of their citizens.

Again, **Materialistic Religions** are **Faith-based** so the followers rely on others for external information and knowledge, which leads to external preaching and external recruitment or ministry of others. In other words, the very **Materialistic followers** just have faith in what they are told and are attracted to the **Order-based leaders** that will just tell them what to have faith in externally. **Order-based leaders** will never teach their followers to look within themselves for the truth through internal communications, but rather will teach them to look externally, to find God somewhere else and to rely on others. This may also seem counterintuitive, but **Order-based leaders** want control and obedience.

On the other hand, **Ethereal Religions** have **Ethereal followers** that want to take ideas and concepts and internalize them. They reject the idea of having faith, but instead develop **Belief-based** systems where there are facts that can be debated internally to test the religious ideas and build up integrity of their religious beliefs.

Of course, **Chaos-based leaders** are attracted to these **Ethereal Religions** because they want and desire followers that will think for themselves and figure it out themselves. Chaos-based leaders are Materialistic and very lazy but they like the status or the appearance of being in charge. They tend to lead Ethereal Religions because the followers are very much into control and so the Chaos-based leaders can be lazy and let them take care of themselves without too much intervention needed.

Ethereal Religions are also Belief-based which demands reliance on each individual follower to study and look internally for knowledge or information and it teaches people to ask questions, seek answers, put forth effort and develop strong beliefs. This creates a religion that sometimes will even shun authority figures but normally will readily accept the lazy Chaos-based leaders. Ethereal Religions also do not accept recruiting or ministering to other people since they believe that each individual must find their own way with their own internal efforts.

Lastly, Materialistic Religions run by Order-based leaders usually have very organized rituals, behaviors and strict membership rules with regular meetings, while the Ethereal Religions run by Chaos-based leaders usually have more relaxed meetings, festivals, meditation and public services.

Religions and today's leaders

As mentioned briefly, today's politicians tend to be <u>Order-based</u>, so they desire a combination religion and political system whereas the state government is worshipped by the people, however their fallback position is <u>Materialistic Religions</u> run by <u>Order-based</u> leaders that control the <u>Materialistic followers</u>, as control and reliance on others is what they desire.

They hate individuals that are self-reliant and would rather have individuals that are dependent on the government or other controlling institutions. They also prefer followers that do not look within for answers or critically think or self-reflect at all. It is the same reason that government-run public education is focused on only the <u>Materialistic</u> side of the human brain, as it helps to keep today's leaders in power.

It is also why <u>Materialistic Religions</u> are the easier ones, where forgiveness can be purchased or supposedly received without much effort, of course through repetitive rituals and routine repetitive prayers.

Going to a church or temple and listening to a leader tell you what the truth is and then having faith that their interpretation is correct makes you <u>Religious</u> but within a <u>Faith-based</u> Religion that is <u>Materialistic</u>. However, if you study a religious idea or thought by yourself through reading and self-reflection and thinking to develop your own understanding and beliefs, then by looking within yourself you become <u>Ethereal</u> and also <u>Religious</u> but within a <u>Belief-based</u> system.

Some of the most popular forms of religion are ones that include both sides of the <u>Eternal Conflict</u>, with both a <u>Materialistic side</u> and an <u>Ethereal side</u> to try and appeal to both types of people. They are also the most controversial ones due to the inherent conflict within the religion that stems from the inherent conflict between both sides. The best religion for humanity would be one that includes both <u>Chaos</u> and <u>Order</u> but tries to balance both against each other.

<u>So why believe in God or even belong to a religion?</u>

There are many people today that proclaim that they are atheistic without having any belief or faith in a Deity at all. While this perfectly fine for <u>Ethereal</u> or

Order-people that see humanity as a whole and strive for the utopian society, I always tell Materialistic or Chaos-people to lie and say that they believe in God even if they don't. The reason is that on the planet Earth, there is a very old Chaos-based idea that humanity should have freedoms, liberties and rights and those come from an understanding of a higher power, such as a God.

Although most people don't understand it, there is a hierarchy of laws that have been established on Earth for all people. The top law that is accepted and comes before all other laws is called God's law and it is the premise that humanity has inherent and unalienable rights that come from a higher power. In the country of America, these are described as the "right to life, liberty and the pursuit of happiness" but again relate to the idea that people have a right to live and to survive and these rights cannot be taken away by other individuals or even governments since they come from a higher power, with the understanding that they are inherent or automatic rights. Of course, this idea is based on Chaos or Materialistic viewpoints, so the individuals that argue against such ideas are usually Ethereal or Order-based people.

The <u>Chaos-driven</u> idea of the first law, which is <u>God's Law</u> upsets many <u>Order-people</u> since it prevents them from taking over or having complete control over other people, of course for the benefit of the complete utopian society. It is also why certain <u>Order-people</u> fight to try and remove God completely from society, from books, schools and even governments.

If God or even religion as a concept can be removed from society, then the first of all laws, <u>God's Law</u> will no longer exist. Then there will be no inherent rights but only privileges given by the governments or leaders and privileges can always be taken away.

<u>The two opposing sides within religion</u>

The phrase "Chaos often breed life while Order breeds habit", by Henry Brooks Adams reveals both sides of each type of religion, while the Yin and Yang philosophy could effectively demonstrate that <u>Order</u> balanced with <u>Chaos</u> are two sides of the same God. (Adams, Henry Brooks C.E. 1907)

Since God is everything and everything is of God, then both <u>Chaos</u> and <u>Order</u> blended together makes up what some refer to as God but could actually be just the entire universe and everything in it. Interesting enough, the Holy Bible is the best-selling book of all time, selling more copies than any other text. I would put forth the idea that the Holy Bible is a collection of books with some being <u>Order-based</u> and some being <u>Chaos-based</u> and the mixture of the two is why this book is the best-selling and revered so much by so many people.

Since the Holy Bible is actually a collection of different books from various religions throughout history, including some that were <u>Order-based</u> and some that were <u>Chaos-based</u> and this is why there are many contradictions to be found within its writing, such as the "Thou shall not kill" statement from the book of Exodus which directly contradicts the later "There is a time to Kill" statement from the book of Ecclesiastes. (The Holy Bible, King James Version, Exodus 20:13 and Ecclesiastes 3:3)

Although some books within the Holy Bible may have different <u>Order</u> and <u>Chaos</u> perspectives, it is generally held that most of the Old Testament is <u>Chaos</u> or <u>Materialistic-driven</u> with a materialistic God that sends forth physical plagues and physical death while demanding physical blood sacrifices from animals and people, while the New Testament is <u>Order</u> or <u>Ethereally-driven</u> with a new covenant that relies on mental prayer, internally introspection and the idea and concept of forgiveness for all.

In fact, the very first morality story within the Holy Bible, which is the story of the Garden of Eden describes one of the first confrontation or conflicts between <u>Chaos</u> vs. <u>Order</u> within the world. The story begins by describing the perfect utopia that <u>Order-people</u> desire, within a garden that has no physical death or physical disease and where everything is ordered with rules and regulations. The main rules given in the story is not to eat of a certain tree, called the tree of knowledge. In the story, <u>Chaos</u> is introduced into this society, in the form of a snake (of course) that offers the ability to make an individual decision, which is seen as breaking the rules. The decision to break the rules and eat of the tree of knowledge is where free-will or <u>Chaos-decision making</u> is introduced to the supposedly perfect world, thereby destroying it. <u>Order</u> arrives in the form of the wind or the winds of God (again of course) which punishes the people by making the Garden of Eden no longer accessible. It is interesting to note that once again, Chaos always takes the form of a snake on the ground the wind.

The two opposing sides create three conditions called Trinity.

C.G. Jung once quoted the Axiom of Maria, with a partial phrase that says, "One becomes two, two becomes three". This quote is attributed to Maria Prophetissa, who was supposedly a third century alchemist and this partial phrase actually has many meanings. Of those many meanings, I personally believe that the symbolism of the number one has always been that of perfection and of singularity. One being the number of God. But if there are two sides of God, then one is called Chaos which is Physical Matter and one is called Order which is Ethereal Energy. That then immediately becomes three relationships between the two sides that are in conflict. (Jung, C.G., C.E. 1944)

Any single entity such as Chaos by itself can only create one condition or relationship, which is the relationship that it has with itself, because there is no other association with anything else. However, if two entities exist such as Chaos and Order then that automatically creates three conditions, which are the two entities in relationship with themselves plus their association or relationship with each other.

Interestingly, the three conditions or relationships created within God have been called many things throughout history and by almost all the religions of the world, but today it is commonly referred to as a _Trinity_ viewpoint of our dual reality, which is usually referred to as God. As discussed above, the _Trinity_ viewpoint of the dualistic universe can be seen within the Holy Bible's story of the creation of _Order_ as the Heavens and then the creation of _Chaos_ as Earth and the subsequent blending of both entities to create everything which is God. However, this viewpoint can also be seen in many other ancient religions that go back thousands of years. In fact, almost all ancient religions agree with the _Trinity_ viewpoint of the dualistic universe which sees the number three as critical to understanding the _Eternal Conflict_.

The ancient Chinese religious symbol of the Yin and Yang, which is that of a circular egg divided into 2 parts and is the symbol of the beginning of time when the physical universe literally emerged from an egg or a round concentration of primordial Chaos. In this mythology, the 1 egg divided into 2 principals from which 3 emerge to create 10,000. Interestingly, this idea that at the beginning of time was just a round egg of Chaos sitting in the cosmic pool of Order and that eventually exploded into the universe as we know it today, is a belief held by almost all ancient cultures and religions.

The ancient Hindu cosmogony called the three relationships by the name of Trimurti, and it includes Chaos or destruction, Order or preservation and lastly, Creation which is the relationship between both Chaos and Order blended together. These three relationships between the two forces were also personified with the entity called Shiva who was Chaos or the destroyer, while the entity called Vishnu was Order or the preserver, and while both entities blend together were called Brahma also known as the Creator God, the association between both Chaos and Order. Their mythology also believed that Chaos started as a cosmic or world egg called Brahmanda, again similar to the Chinese story.

Within the ancient religion of Buddhism is the same trinity idea but is called the Three Jewels, and it includes Chaos as the Dharma, which are direct visible experiences and physical sufferings, Order as Sangha, which is the one true way that never changes and finally the Buddha, who is the blending of both elements resulting in perfect enlightenment and true knowledge.

The ancient Egyptians referred to their Trinity as the entity called Sekhart who was Chaos and the entity called Ptah who was Order while both entities blended together were called Nefertum. The ancient Egyptians also used the Trinity viewpoint within their Isis, Osiris and Horus mythology.

Classic antiquity included the idea of the Goddess Diana as a blending of three and which included the Goddess of the Nether world, the Goddess of the Moon and the blended Perfect Huntress, while the ancient Roman mythology includes the Three Fates or even the Three Furies. Even stories relating to famous events or creatures almost always included three, as in the three heads of the famous Cerberus or even the great pyramids of Giza, Egypt being three in number.

Today, the Trinity viewpoint can also be seen in most Christian religions, through the entity of Jesus, the physical son of God as Chaos and whose death brought freedom to humanity and the entity of energy called the Holy Spirit as <u>Order</u> who brings information and knowledge to humanity, while both entities blend together as God called the Father and who is known as the creator and also a part of everything, including the association of the first two.

Almost all religions and philosophies, both ancient and modern look to the number three as a relationship between the two major powers as simply a given fact, sometimes not even bothering to explain the overall idea. Even modern-day psychology still accepts the ancient views of the Greek philosopher Plato who described all humans as having a <u>Tripartite Soul</u>, where the physical body was full of physical desires and appetites, the invisible soul which was made up of spirit and will and lastly, there is the relationship that both physical and spirit had with each other. (Plato, B.C.E. 380).

This belief in the <u>Tripartite Soul</u> within humanity and within the universe is one of the reasons that the Greek philosophers and mathematicians developed ideas revolving around triangles, which have three sides. Pythagoras, one of the most famous Greek mathematicians, even taught his followers that everything in the universe had a three-part structure and that you can understand every idea and even every problem, if it is broken down into three parts, like a triangle. By understanding the two forces within the universe, which are Chaos and Order and their relationship to each other, then it is easy to understand this viewpoint that all life can be understood and broken down into three parts or three relationships.

While most ancient and even modern religions and mythologies include a trinity, tripartite, triune or triadic viewpoint regarding their deity and even the universe, most also include at one individual that had three personalities or even three heads, all representing the trinity of Chaos, Order and their blended relationship. Some ancient mythologies would also include individuals with one head but having two faces exactly opposite each other, representing the same idea. In fact, Carl Jung was so fascinated with the arrangement of deities into groups of three that he considered this to be a major archetype of all religions throughout history.

Religions can become perverted with the two E's

There are many variations of religions, but as you can see they all have concentrations of Order and Chaos.

Both types of religions are important and needed, mainly because there are both types of people within the world and in fact some religions try and blend both sides to form a universal religion that appeals to all people. There is no problem with either type of religion, especially if they help people or they help to form a basis for self-identity or the identity of a society. Religion can become a good habit to help nourish either the <u>Physical Matter</u> side or the <u>Ethereal Energy</u> side of each individual, but only if done correctly with the need for balance. There are however, some religions that are perverse or have become perverse and you can spot them easily if you understand the two E's, which are <u>Exclusive</u> or <u>Extreme</u>.

The first perversion of religion is that of being exclusive. The idea of a religion being exclusive is the idea where God is difficult to reach or access, unless of course you belong to this specific religion. Belonging to this specific form of religion is that only way to be saved or to access God. This creates the exclusionary attraction, which means that if you belong to that specific religion then you are special. That God is only available to you and you only, since you are now one of the selected, special or chosen people that God loves.

This then extends to the idea that God is not available to everybody, but only just to this special group within this special religion. This also may even extend to specific religious teachings that all the other religions are wrong and should be hated or despised, especially because only you and your specific religion have the elements of God correct.

The second perversion of religion is that of being extreme. Extreme religions take the concept of exclusivity even further by implying that God accepts only you and your specific method or system or worship, which also means that God has now purposefully excluded all others. Extreme religions usually teach that God is not love but rather that God is angry and violent and demanding because of the other groups that have now rejected God. People that belong to extreme religions can easily become <u>Religious Fanatics</u>, which in its final form have led to most of the historically terrible wars of the ancient world and even some of the modern ones. Most of these religious wars have all been started because of the <u>My God is better than your God</u> mentality.

<u>Religious Fanaticism</u> always starts from <u>Materialistic Religions</u> that are <u>Faith-based</u>, that become

exclusive and eventually extreme. The exclusionary teachings usually focus on the idea that only select followers of the specific religion are chosen or special which leads to the understanding that all other people have no value or importance and should therefore be eliminated or killed. This idea is based on the teaching that God only cares for the true followers and not anybody else.

This exclusionary extreme idea feeds the <u>Id side</u> of the human brain, which is the animalistic <u>Materialistic-side</u>, and which eventually leads to the followers being prevented from even questioning or thinking about reality. Hence <u>Religious Fanatics</u> are born that have only faith in a strange concept, whereas they are the only ones that have the true religion, while everybody else is wrong and therefore must be eliminated.

Even though <u>Religious Fanatics</u> are very dangerous and create disasters, death, wars and confusion - they still exist, and it is because of the purposeful division that occurs. The more the world populace stays divided, isolated and confused, the easier it is for leaders to control others. With this is mind, anytime you hear politicians or religious leaders giving speeches that separate or divide people based on religion, race, sex or any other concept, then you should immediately understand that this method is being used to prevent the followers from uniting.

Chapter 6 - TYPES OF GOVERNMENTAL SYSTEMS

The different forms of Governments based on Order and Chaos

Today, you hear many different people talk about different types of governments or types of governmental officers. Like the duality of the universe, the descriptions are usually two sided, such as left-winged or right-winged. Many other descriptions revolve around <u>liberal leaders verses the conservative leaders</u> or <u>small-government people verses big-government people</u>.

Now understand that all these descriptions are all wrong and are made once again to distract the people from the truth. As described earlier, this incorrect description is being done like the <u>good against evil</u> argument that is also made incorrectly and no purpose. To really understand the types of governments available and the type of governmental leaders that exist, we must once again turn to the <u>Eternal Conflict</u> and view governments within the model of <u>Chaos</u> and that of <u>Order</u>.

The first of the actual governmental structure is one that is made up of all <u>Chaos</u> without any rules or regulations, but rather has only freedoms, rights and liberties and this form of government has been called Anarchy. The problem with Anarchy as a governmental structure however, is that it is based only on the materialistic physical realm or truly Complete Chaos, so it never really exists for very long at all. Anarchy exists for only a moment until the freedoms of some people directly conflict with the freedoms of others and then criminal behavior happens. The excess amount of Chaos, without any balance, will quickly destroy this simple governmental structure called Anarchy, which is really the absence of any governmental structure. Therefore, Anarchy is actually dismissed as a legitimate form of government because it never lasts very long.

    On the other end of the spectrum is the governmental structure that is made up of all Order without any freedoms, rights or liberties, but rather has only rules and regulations and is called a Dictatorship. The problem with a Dictatorship, as a governmental structure however, is that all the power is concentrated by one individual, so that Complete Order exists for that one individual, which quickly becomes the Addiction to Power. The excess amount of Order, again without any balance, will quickly led to a revolution by the mass of followers and the governmental structure called a Dictatorship, like the quick outcome of an Anarchy government, it will quickly be eliminated. Also, a Dictatorship is never an actual reality as one person is never really given all the power and so it actually never exists. Usually it is a leader with a small group around them and it is the small group with the leader that has all the power. This form of government, with a small group in charge, even if one person is the figure-head is actually called an Oligarchy, so a Dictatorship as a government, again similar to an Anarchy government, is usually dismissed as an actual legitimate form of government, because it never really exists.

So, we can dismiss the two extreme ends of the spectrum of government, one which is Complete Order as a Dictatorship and its opposite which is Complete Chaos as Anarchy, because neither exist for very long at all and both are illegitimate forms of governments. So, that leaves only three legitimate forms of government that can exist and are located somewhere between Complete Chaos and Complete Order. These are a Republic form of government, a Democracy form of government and an Oligarchy form of government. We will now discuss each of these three forms of government, again within the context of the **Eternal Conflict**.

A **Republic** form of government is one that is defined by the **Rule of Law** but with a limited set of laws (usually lawful laws based around a Constitution) and the axiom to lead by the rule of these limited laws. **Chaos-People** create and love **Republic** forms of government as they allow people to be completely self-sufficient while keeping the government small. They understand the need for some simple basic laws and will insanely protect these few laws as much as necessary. **Republics** are normally created with the understanding that the first major task of government is to simply protect the security of the people while also protecting the freedoms, rights and liberties of the people. Since a **Republic** is a **Chaos-ba**sed government setup under a set of very limited laws and not based on a majority vote, a **Republic** is the only form of government that really protects any minority group. **Republics** are usually out-of-balance however, due to them being heavily weighted toward **Chaos**, which is really the main reason that **Republics** historically do not last very long.

A _Democracy_ form of government is one that is defined by a vote of the majority. If you have 51% of the vote, then you get anything you want. Historically however, a _Democracy_ is a transitional form of government that usually takes over and moves away from a _Republic_ and it usually starts with the _Chaos-based_ lawful law systems being replaced by the _Order-based_ legal law systems. This happens however, because of the out-of-balance condition of too much _Chaos_ and so more and more _Order_ and control starts to get added. Once rich and powerful people within any society, realize that they can buy 51% of the votes, they can then achieve a majority and then they can force the government to give them anything. It is at this point that this _Democratic_ form of government will start to move away from a _Republic_ form of government and will eventually become an _Oligarchy_.

Again, _Democracy_ forms of government are usually terrible for any minority group as they are quickly silenced by the majority voters or the purchased 51% of the vote. _Democracy_ forms of government as also the ones with the most conflict and the most problems as _Order-People_ will slowly start to move the _Democracy_ toward absolute control, which again is an _Oligarchy_ form of government, while Chaos-People are trying to regain their freedoms and are trying to move the Democracy back to a Republic form of government.

An Oligarchy form of government is one that is defined by a small group of people running everything, usually with a figure-head that is the perceived leader, even though that person may not actually have any power. This is the form of Government most loved by Order-People as all individuals, businesses and society in general are under the most control with the most rules and regulations. Oligarchies are usually out-of-balance however, due to the being heavily weighted toward Order, which is the main reason that true Oligarchies historically come to an end with violent revolutions.

Interesting enough, if you follow governmental history throughout the world, you will find that Order-based Oligarchies will eventually be overthrown, because the people end up feeling completely suppressed with no freedoms and then eventually a revolution of those people will happen. This revolution will start as Anarchy for a few days until some version of a Chaos-based Republic is created. Once again however due to lack of balance, the Republic will be transitioned into a Democracy with lots of problems and conflict with both <u>Chaos-people</u> and <u>Order-people</u> fighting each other. Unless the people have weapons or a means to fight, usually the <u>Democracy</u> always becomes an <u>Oligarchy</u> when the government uses force and the cycle starts over again. Therefore, some form of freedom to keep and the citizens must be allowed to maintain weapons for self-protection are necessary, especially to prevent <u>Oligarchy</u> types of governments from happening. Any government official that demands gun-control or the taking of arms from its citizens is only after absolute control and rule through an <u>Oligarchy</u>.

This cycle of governmental changes from <u>Republic</u> to <u>Democracy</u> to <u>Oligarchy</u> and back will continue forever and has throughout world history until everyone, both <u>Order-people</u> and <u>Chaos-people</u> realize the need for balance between the two forces at work within the <u>Eternal Conflict</u>. Both kinds of people must get together and work toward creating a government created out of balance for true governmental conflict to stop.

<u>The Eternal Conflict forms your opinion of governments.</u>

<u>Order-People</u> believe in rules, regulations, statutes and control. They are not bad people, but rather

in their hearts and minds they believe that society or people at large cannot and will not take care of themselves. They believe that individual people do not have unlimited potential and power but are instead chaotic physical animals that must be controlled. Therefore, they believe that the best form of government is an <u>Oligarchy</u> where the government is a nanny state or a police state that is directly responsible to take care of and control the population with rules and regulations and force if needed, because they view people as physical creatures that make terrible chaotic decisions.

<u>Order-People</u> believe in the very limited potential of people, so they fight against allowing people to become self-sufficient or even to allow them self-protection or weapons as they truly believe the population in general is not capable of being self-sufficient or of even protecting themselves. This tendency causes <u>Order-People</u> to be viewed by others, especially <u>Chaos-People</u> as psychopaths or sociopaths, although they do not see themselves in the same way. <u>Order-People</u> see what they would describe as the big picture of an overpopulated world of poor limited people that cannot take care of themselves and that creates a

vision of them having to make the hard decisions for everybody else.

The <u>Order-People</u> call themselves progressives, liberals, socialists or communists and again they are not bad people but instead they are people that believe that a government with lots of control is the answer and that government must provide for everyone, because most people are too stupid, lazy and driven by <u>Chaos</u>. This is a weird belief but honestly can be held up to reality, because people do form habits and addictions and will be lazy and stupid if they can be. If the government will send free money to an average individual to sit at home and do nothing, then most people will comply and take the free money.

<u>Chaos-People</u> are opposite and truly believe in the unlimited potential within each individual, so they want to give each individual the choice to do whatever they want. They believe that people can take care of themselves and become self-sufficient but only if they are given a limited government and limited rules and regulations in which to perform. In some instances, they will even give people the right to not be ruled or not to consent to be governed, which is called a <u>Free Man of the</u>

Land. They also understand human nature and so they know that most people must be forced into action, but that once they are in action then each individual can become self-sufficient and can use self-protection. The problem with this belief is that too much freedom is too much Chaos and creates social problems and crime. So once again, moderation and balance are necessary.

Each viewpoint is valid, as if you look around the world today, ask yourself just how many people argue or throw temper tantrums? How many people ask for licenses or register things with the state, which is just asking for permission like children do? How many people want to be left alone and just taken care of by somebody else? If you answer these questions honestly, then you will understand that there is a need for balance that must happen, or the destructive pattern of history will just continue to repeat itself.

It is interesting to note that Order is not only control but control of invisible things like ideas and thoughts. Order-People are usually the first types of people to want to censor books, newspapers or even the internet. Order-People believe and desire limiting speech and especially disagreement, whereas Chaos-People believe in the freedom to speak out and they usually disagree with any kind of censorship. The **Eternal Conflict** extends throughout the various governments today and even around the world. If you pay attention, then you will notice that when different governments interact and especially when there are problems between governments, the **Order-People** will always suggest dialog, sanctions, treaties or verbal international condemnation, whereas **Chaos-People** usually jump to physical threats, physical attacks, physical wars and even justified physical death.

Order-People are the first to suggest, desire and add laws, rules, regulations and statutes. In reality, they do not believe there can be too many laws as they believe that everything must be regulated and controlled. They are the ones that bring Legal laws to the population, which are made-up laws that govern ideas only.

Chaos-People on the other hand only want a few basis laws that are based on physical harm or physical damage. If no physical person, animal or object was harmed or damaged, then they do not believe any problem occurred. This is the basis for lawful laws.

Order-People also understand that humans in general have three traits; pleasure seeking, pain avoidance and only expending as little energy as possible to survive. So, they create Order through control by demanding that the government always provide these three things for all people without expecting effort and labor in return. The problem again is that this creates an out-of-balance situation whereas any country will then eventually fall into a slavery or serfdom system where the government must force labor upon people or the country becomes bankrupt. Again, moderation and balance between the two forces must happen.

Government types affect other systems.

Around the world today, there are those people that support Chaos around the world and there are the opposite viewpoints of those people supporting Order. Again, the various people may not even be aware of why they support each system and may not even see the two different systems as separate or real.

Again, governments that are mainly based on Chaos will also extend Chaos into all other systems and all other areas. For example, a Republic form of government will be limited with limited rules and regulations that allow people to make their own chaotic decisions with the belief in unlimited potential. However, this viewpoint also allows for unlimited failure without a social safety net to prevent their failure. The problem with this viewpoint is that even though certain highly motivated people can possibly reach their unlimited potential, thereby becoming incredibly rich with huge amounts of power as an individual, then other individuals will fail miserably and not have a social safety net to help them.

Again, the opposite type of governments based on <u>Order</u>, such as an <u>Oligarchy</u>, will try to take <u>Chaotic</u> decision-making away, so they are always adding more rules and regulations that limit or even prevent freedoms, rights and liberties. Again, <u>Order-People</u> cannot have enough law, rules, regulation and statues. All the rules are put in place to mainly prevent people from failing because again, <u>Order-People</u> truly believe that the majority of people have extremely limited potential and so a very extensive social safety net must be in place for the government to literally take care of people as they cannot take care of themselves. Of course, the problem with this viewpoint is that it literally prevents highly motivated people from reaching their actual potential and it also limits any motivational level for the average person. Why work hard if the benefits that you will receive are the same for everyone, whether you are lazy or not?

These two opposite views inside any governmental structure will also extend into business, energy, banking and all other systems through laws, rules and regulations based on Chaos or Order. For Republic forms of governments created by Chaos-People, businesses will be mainly divorced from government with little or no backstop to prevent failure. The view is that businesses are the same as people and will succeed or fail based on their business model and the market forces, keeping in mind that market forces are also Chaotic and difficult to track. However, businesses within a Republic are really the only ones that have the possibility to reach their unlimited potential, thereby creating huge rich businesses with incredible amounts of money and therefore power.

Oligarchy forms of governments created by Order-People do not allow businesses to become powerful as they will be controlled by governments and may even be part of the government system itself. Some businesses that are deemed critical to the state will even be prevented from failing through regulations and supported by taxes but also prevented from becoming too large. A side effect of this **Order** and control system of government is that it stops or even unknowingly prevents new ideas or new intuitive businesses from replacing the old systems, similar to what happens with individuals.

Going hand-and-hand with businesses are the types of laws created by the different forms of governments. **Lawful Laws** also called the **Law of the Land** or **Common Law** are created within **Republics** and these lawful laws will be based on physical damage to actual property or real physical harm to real physical things. Breaking any lawful law means damaging or harming real property or physical things or people. Also, businesses will have unlimited potential but also unlimited liability under these laws because they are considered real physical businesses.

Legal Laws are created under **Oligarchy** forms of governments, which are also called the **Laws of Commerce**. Since these laws are based on **Order**, they are made-up and based on invisible fictional corporations that have only limited liability. Breaking any **Legal Law** means simply not following a made-up statue, rule or regulation. Harm or damage to real property or real people is not important. Sometimes just the potential to damage can be punished. For example, driving too fast or not purchasing a license is a violation of legal laws. Legal laws are all about control, regulation and limiting freedoms, rights and liberties. There will be more discussion about the different laws and two different law systems based on the Eternal Conflict in an upcoming chapter.

Republic forms of governments will also allow people to make decisions regarding self-sufficiency based on energy needs. Chaos-People believe that all people should fail or succeed based on their own decisions, so the three basic critical needs, that being food, water and shelter will not be regulated or controlled. Hunting, fishing, farming, drilling water wells and building homes will be allowed without much government involvement. Also, experimentation with free energy systems or alternative energy systems may even be encouraged.

Oligarchy forms of governments do not allow people to become self-sufficient as this will lead to unfairness, with some people not needing even the government while others are completely dependent upon the government. Hunting, fishing and farming will be over-regulated with licenses and registration requirements to maintain the wildlife habitat. Also, homes or property are sometimes not even allowed to be owned by the people, so property taxes will be levied to prevent self-sufficiency and the unfairness of some people to own their own shelter while other people don't or can't.

In regard to property tax, it should be noted that Order-People will say that property taxes are always needed to pay for schools and education, but the reality is that other forms of taxes could be raised for those educational programs. Property tax as a concept comes only from Order-People to prevent self-sufficiency because even if you own your own house with no debt or mortgage, if you then stop paying the property taxes, then the government will simply take it from you. Also, experimentation with free energy or alternative energy systems will not be allowed within Oligarchy forms of government, with some even being against the law in most circumstances.

Republic forms of governments always form money systems based around Materialistic valuable physical objects, such as gold or silver-based currencies. This creates money that has both the <u>form of money</u> and the <u>substance of money</u> but also limits the money supply of a nation based on the availability of the physical substance. A limited money supply also helps to limit the size and scope of government and prevent the government from over regulating. This also helps people that make good decisions and allows for the use of money without any government interference as the government cannot track commerce based on this.

<u>Oligarchy</u> forms of government will create money that only has perceived value with just the <u>form of money</u>, such as notes, fiat paper or debt-based currencies. This is done to help the government control the money supply with serial number-based paper script. The government based on <u>Order</u> will also allow for fractional reserve banking systems that provide unlimited growth of the money system completely within the control of the government. Unlimited growth is what is needed to also the <u>Order</u> based government to also grow larger and larger with more and more control. Two different types of Money systems also based on the <u>Eternal Conflict</u> will also be covered in the next chapter.

The pattern of Government types based on The Eternal Conflict.

As mentioned earlier, since <u>Anarchy</u> and <u>Dictatorship</u> forms of governments never really exist except for brief moments, there are only three legitimate forms of governments and they are <u>Republic</u>, <u>Democracies</u> and <u>Oligarchies</u>.

Although we discussed the transition from governmental structures earlier, let's go into a little bit more detail, because as you track world history, you will discover that these three legitimate forms of government change constantly within different countries and within different societies. The constant change of governmental structure has always been a part of humanity of Earth and is based once again on the <u>Eternal Conflict</u> and the out-of-balance situations that have become inherent within the duality that is life. Although the change in governmental structure, when they occur, is <u>Chaotic</u> and sometimes destructive, there is a certain pattern to the movement and change. Take any society that starts out with <u>Complete Chaos</u> without any <u>Order</u>, therefore zero balance between the two.

This society will quickly make some rules and they may start out as unwritten social norms but will also quickly be turned into written rules based on limited scope. The classic example is found within the Holy Bible when the Israelites were no longer slaves in Egypt and they had instantly become free people that were wandering as a chaotic group in the desert. Moses, supposedly under the direction of God, quickly brought the people a set of small limited rules called the Ten Commandments. (The Holy Bible, King James Version, Exodus 20)

Once a specific set of rules are created, the governmental form becomes that of a Republic which is a Rule by Law which is limited and run by the people themselves. This form of government allows for unlimited success of the people but the potential for unlimited failure. Some classic examples of this historical pattern occur within the Ancient Grecian Republic, Ancient Roman Republic or even the modern American Republic that formed in the late 18th century that has already transitioned over to a Democracy today. Within any Republic is the inherent lack of balance with too much Chaos, which becomes apparent with people that will take advantage of the lack of rules and form monopolies and quickly gain power and control. Sometimes these people and their businesses will end up with more power and control then the actual government itself. This is just the Eternal Conflict once again trying to find its balance.

So, as people or businesses become more and more powerful with more and more control, then they will start to change the systems around them. Soon, money systems will move away from lawful money based on physical gold or physical silver and move to legal tender based on perceived value only. Also, lawful laws based on real property will be supplemented by legal laws to protect the rich and powerful. This transitioning phase that moves away from Chaos-directed Republic forms of government over to Order-directed Oligarchy forms of government is called a temporary Democracy form of government.

There have been many leaders throughout history that have called a Democracy the worst form of government ever and in some respect, they are absolutely correct. The main reason a Democracy is in reality a terrible government, because it is always transitioning and changing the landscape and also has with most people hating any form of change. A Democracy always moves away from the Republic with too much Chaos and moves toward an

Oligarchy that always has too much Order, tipping the scales one way and then the other way.

At first the Democracy form of government happens to a Republic without people becoming aware of the change directly as the change is slow, but soon people will suddenly become aware that they have less and less freedom and fewer and fewer rights and liberties. Suddenly their limited government called a Republic is disappearing and the government is becoming too powerful and too large as Chaos becomes replaced with Order.

Second, a Democracy adds massive amounts of taxes and fines and fees. This happens because as the governmental structure grows and becomes huge, somebody must pay for the added governmental workers and enforcement of the massive amounts of added rules. Sometimes, the government will borrow the money needed to create faster change without the people becoming aware of it.

Third, a Democracy becomes the worst form of government once people realize that the Republic, which was the Rule of Law is no longer being applied. Suddenly there is a realization that there are two sets of laws, lawful and legal, and that the rich and powerful use a different set of laws then the regular people.

Lastly, as the amount of Order takes over and Chaos is suppressed, even the Democracy starts to fade away. That happens once the rich and powerful realize that they only need 51% of the vote to decide anything, including voting themselves more money from the government. So, rigged elections and suppression of voters happen which results is disillusionment of the people. Worse, any rights that minorities had will start to be eliminated as they never had 51% or any majority vote and the Republic which was the only form of government that protected them is now gone, as it is disappearing fast.

The government is now transitioning to an Oligarchy which means that it is governing by only a small group of people and they will become so powerful that any group, such as unions, charities and even religions, can and will be eliminated by third small group of people which is now the government. The people will now find themselves without any freedoms, rights or liberties as Order is now too strong. A revolution and much bloodshed will eventually be the result as Chaos will fight back.

The **Eternal Conflict** is real and once again, within both real forces is a need and desire for balance. Although both sides are in opposition and conflict, they also have the attraction to each other since opposites attract. The best form of government is one that balances both sides and has major checks in place to prevent either **Order** or **Chaos** from becoming too strong.

Keep in mind that the best in-balance government will protect its people, their employment and the prosperity of the country but not at the expense of other nations. The best government wants self-sufficient citizens and works hard to protect the people and their rights. Today the word protectionism is often seen as a bad word, but it is not necessary the truth. Both internal and external stability and especially balance between **Chaos** and **Order**, within any nation and also within the world, are the keys.

The next chapter will expand on these ideas, by further explaining the money systems used under the **Eternal Conflict**.

Chapter 7 - DIFFERENT SYSTEMS OF MONEY

The different types of cultures based on the Eternal Conflict.

Like the different types of religions or the different types of governments based around the Eternal Conflict, there are also two different types of cultures that form on Earth and they are also based on the Eternal Conflict. It is important to first understand these two different types of cultures in order to understand the two different systems of money that can arise within each type.

The first few types of ancient cultures known to humanity were based on the Physical Realm of Chaos, started as a Hunter Society, an Agricultural Society or as a Herder Society. All three of the different types were all based on physical things, such as animals to hunt, crops to plant or domestic animals to raise for food, but the Hunter Society had a tendency to be nomadic in nature as they moved to follow the animals that they hunted, while both the Agricultural Society and the Herder Society were very stationary and protective of the physical land that they farmed or let their animal graze on. All the societies based on Chaos or physical things eventually started to develop different selection methods of picking the best animals or the best seeds to create better physical things. This idea of genetics and selection was the start of the culture moving in the direction of Order through control and selection.

Eventually however, these societies that start out based on physical things are again out-of-balance with too much Chaos, so the societies will eventually build cities and organizations based around more extensive ideas and concepts, as ideas and concepts are invisible parts of the Ethereal Energy system within the human brain as a part of Order. Today, we have cities in America such as Manhattan, New York and areas such as Silicone Valley, California that have transitioned almost completely to a society based entirely of Order. This is because one of these ideas and concepts is that of Money and the idea of control through money, banking and financial arrangements.

But let's back up a second, because as societies start to move away from Chaos and toward Order, they will eventually become an Industrial Society based on new ideas and inventions that need less physical labor to make the same foods or products. **Industrial Societies** will also create and use new materials and metals as the machines become more complex as the ideas become more complex. Interestingly, the **Industrial Society** usually never forms until the idea of **Money** takes root and is accepted by most humans within the culture. An **Industrial Society** is the transitioning society away from the **Physical Realm** of **Chaos** toward the **Ethereal Energy Realm** of **Order**.

The next society in the transition is that of almost pure **Order** or the **Informational Society** where ideas, thoughts and information become the most important product and service. In this society invisible information can become so valuable that it can be sold or exchanged for another invisible idea or concept called fiat money. However, much like governmental systems or religious systems, the **Informational Society** will quickly become out-of-balance with too much **Order** and not enough **Chaos**, as the need for information becomes the norm and the freedoms, rights and liberties of the people almost disappear while surveillance and information gathering grows to become overwhelming to the individuals living there.

Privacy and the rights of people as a part of Chaos will almost disappear completely within this society as Order becomes too large and unwieldy. It is during these times of overwhelming government interference and demands that you will also see terrorist groups based on Chaos rise up that want to destroy this Information Society that has swung itself completely out-of-balance with Order controlling everything. The creation of terrorist and terrorist groups today is once again, just the two forces of Chaos and Order simply trying to reach balance.

The introduction of money to Chaos-based cultures.

The Chaos-based cultures, which are based on physical animals or physical crops almost always start out as totally self-sufficient societies. In other words, to survive the physical trials of physical life, these cultures must obtain three things; finding shelter, maintaining a food supply and maintaining a water supply. These are the three basic necessities of life and must be obtained to prevent death from occurring in humans.

Also, these strictly Chaos-based cultures usually have no need for money as we know it today. In fact, they usually don't even understand the concept of money or even the need for it, as they usually barter or exchange physical items with each other to insure the success of each individual and survival of the social group. Money however is a critical idea for Order-People because it is a crucial tool that is used for control.

Whenever a group of Order-People will come into a Chaos-based culture, they come with different ideas and thoughts. The main idea is to try and eliminate bartering and also to create a system of trading, mainly of things that are different from the three necessities of life. Trade to them is like bartering but usually involves trading for something non-essential, like jewelry, unique food items, drugs or alcohol. They do this by appealing to the normal habits and addictions that all human have, especially the Chaos-People by first giving away the items as free samples and then by bartering or trading for them later, once they become desired.

Over time, the desire for these new products that are not necessities will grow strong within the society of Chaos-People, which is when the traders suddenly stop bartering and instead insist that the goods and products be purchased. This new concept called purchasing requires a new concept called money, which means that money as an idea will be introduced next. This leads to more <u>Order</u> being added to the out-of-balance <u>Chaos</u> that was in the society and helps create balance, so it is usually not fought against that much.

Money introduced to <u>Chaos-based</u> cultures will always start out as physical items, such as grain, silver or gold. The first money will always be that of <u>Lawful Money</u> with the definition of <u>Lawful Money</u> being something that has both the actual physical form of money and one that has actual physical value or worth. One example of <u>Lawful Money</u> can be gold or silver coins.

Over time however, the <u>Order-People</u> or traders will slowly change this concept of money into that of <u>Legal Tender</u>, which is imaginary money that although it may still have an accepted physical form of money, it will no longer has physical value. An example of this will be a piece of paper money that simply has different numbers or different values printed on them. Eventually the <u>Order-People</u> called bankers or traders, will over time try to eliminate the actual physical form of money, so that there are only just imaginary digits displayed on a computer screen. By eliminating money's physical value and physical form, money will end up just being <u>Complete</u>

Order by simply being invisible energy or data stored in a computer.

The introduction of money as a commodity.

The introduction and eventual acceptance of money happens when the Order-based traders introduce money to the Chaos-based physical society gradually, by first taking a physical commodity that the society understands, such as copper, gold or silver and they start coining that commodity.

Take note that the word coin is actually a verb and not a noun. The process of coining means to measure, weigh, size and then stamp out the commodity. For example, some silver is weighed and exactly one ounce of silver is set aside. This one ounce is then measured and found to contain the correct purity or number of grains of actual silver. It is then sized into a flat round object, square cube or rectangular bar and then stamped with words that say something like, measured by this country to be exactly one ounce of 0.99 pure silver.

The reason that the traders start with coining physical commodities is that this concept can easily be understood by Chaos-People, as the physical commodity is readily understood to be physical money and this physical money will simply be inserted into the normal barter system that was being used. But instead of trading one chicken for two pieces of fire wood, they can now trade one chicken for one silver coined piece and then trade the silver coined piece for two pieces of fire wood.

If the Order-based traders cannot get the people to accept this strange new concept called money, then they will usually put one of them in place as a leader, infiltrate the government to get acceptance or try to bribe the leader(s) by putting them in charge of the money creation process which instantly creates a lot of power for that leader. The leader will then help force the money into circulation to maintain their temporary power.

Remember that Order-People are traders or bankers only because they want and desire Order and control. They see money as a means to get to that end result very efficiently. Lawful Money or physical money based on commodities are not a bad thing, if it is left alone at that point. Physical money can be extremely beneficial to the group, because it helps to facilitate trade with other nations using a common currency. However, there are four rules that must be followed regarding Lawful Money to help balance both Chaos and Order.

The first rule to maintain balance is that money must be made of something physical, such as wood, steel, diamond, gold or whatever. But it must be physical and must have value to the society. They must never create paper or digital money.

The second rule is that only the government or leaders can issue any money. It must be created and then spent by the leaders into circulation. Never allow the creation of money to be privatized to a private banker or someone other than the government.

The third rule goes in hand with the second and that is that the amount of money available must be tightly controlled to meet only the needs of the society and the total amount available must not be decided by the government that issues the money, because they will then always issue too much. The total amount to be issued must be controlled by a third party apart from the government or leaders to prevent inflation or deflation. This of course is the tough one because controlling the amount of money also limits the potential expansion of the society. In fact, <u>Order-People</u> use the idea of unlimited expansion as one of the main reasons for a country to give up their <u>Lawful Money</u> and switch to <u>Legal Tender</u>, which is imaginary money based on <u>Order</u>.

The fourth rule is that money must never be created as debt with attached interest owned. This is critical to understand, because debt-based money, such as the current U.S. debt-based Federal Reserve Notes, get created no means of ever paying the debt back because the future interest that must be paid on it is never created. This type of money just enslaves people, because as was just mentioned, only the principal is created and the money that is paid back as interest must come from someone else and that person will eventually be financially destroyed to make this happen. It is the reason that debtor's prisons are eliminated and <u>rules about bankruptcy</u> must always be in place when debt-based money is used within a society because as the rich get richer, the poor must suffer. Mainly because as the interest is paid back then the money supply for the entire society shrinks and the poor end up going without.

Once the traders introduce <u>Lawful Money</u> based on <u>Chaos</u>, they will, as mentioned earlier, slowly try to change it into <u>Order-based</u> money called <u>Legal Tender</u>, which is based on debt or negative numbers. All of this is want <u>Order-People</u>, as traders and bankers desire, which is to control all people through debt servitude, especially debt servitude that people are unaware of.

<u>The number zero and the introduction of money as debt.</u>

Interestingly, for debt or negative money to come into existence, the <u>Order-People</u> must introduce a strange idea call the number zero. Today we readily accept the idea of a symbol or number for nothing, which is zero. It is based on our current mathematics models and is taught to all students inside the current government-run school systems.

However, most of the early <u>Chaos</u> civilizations based on physical things never had a symbol for zero or nothing. In fact, most of our ancient civilizations would not even understand the need to identity nothing. Again, the ancient cultures that were <u>Chaos-based</u> such as hunting, agricultural or herding animals did not have even a concept of the number zero. You either had something or you didn't have something. Why would you need a symbol or number or word for something less than you have or possess?

We may not see it today, but the idea of zero or even negative numbers is a very foreign concept to any ancient <u>Chaos-based</u> group that succeeds at self-sufficiency. In those groups, you only use what you physically have to barter or to trade. Once negative numbers or debt can be understood by an ancient culture, then the next concept of debt or borrowing money can also be introduced. This idea of debt again is critical for the <u>Order-People</u> to gain control, so what happens through the traders and the bankers when they flood a society with products and services that the people did not even know they wanted, such as different forms of food, jewelry or unique products.

Most times, the flood of products or services will almost always start out at a very low price, sometimes even being provided under the cost of making the product. This creates a desire and need for the product and its cheap cost makes it readily available to everybody - at least at first. For example, the <u>Order-People</u> as traders will sell you corn for cheaper than you could grow and harvest it yourself, so the group will eventually stop growing and producing corn and instead start buying it. Also, the traders will purchase certain products from you at a very high price, so that the society as a group can be tricked into growing only one product that they can make lots of money off, and this tricks the society into not being self-sufficient anymore. The fewer products that the group makes, the more products the group will have to eventually purchase from the traders.

If you look through history, you will discover that same pattern of only certain products being made in certain countries with cooking oil coming from the Middle East, coffee beans from South America, cotton from North America, etc. Again, <u>Order-People</u> cannot control groups that are self-sufficient, so tactics and tricks are employed to make each society dependent upon the traders. Sometimes, they will even manipulate the governments themselves so that laws will be passed to prevent people from producing certain products, especially the products that are necessary for survival or self-sufficiency. One such example from modern times was when the people of India broke the British Law when they became self-sufficient by simply making their own salt as shown by Mahatma Gandhi.

Once these systems are in place, then the final step happens, where the trader starts to slowly raise the prices of the products that they sell and slowly lowers the prices of the products that they buy, so that eventually there comes a time when the group finds themselves unable to purchase the products that they need and they are no longer producing those products themselves. Then the concept of borrowing money and debt becomes a reality for the different societies. These societies that were based on physical things, suddenly find themselves needing to borrow money in order to survive. If you look to history, you can also find that the Order-People as bankers or traders will sometimes create wars between nations, behind the scenes. That way they can step in to borrow the same nation's money or credit to finance the war and Order-People, as bankers, will sometimes even finance both sides of the war thereby allowing them to control both nations eventually.

The concept of taxes by governments

Once these systems are in place and the societies start to find themselves in debt, then the concept of taxes happens. Taxes are money or wealth taken from the people and given to the government to allow it to run.

Self-governing societies that are set up by <u>Chaos-People</u> usually have very small governments and therefore do not need to impose taxes on the people or if they do, the taxes are very small and manageable. Keep in mind that all governments do not make a product or service that people need, except for basic necessities such as to protect the people with fire fighters or police officers or to protect the rights of the people through lawful judicial systems. Everything else that governments does is the government deciding that they should do it, adding more governmental offices and people and then taxing the nation to do what they want to do. Remember that the more money and taxes that are taken from the people, the less that the people must spend in the normal economy.

If the <u>Order-People</u> are allowed to continue then even the governments, which never need to borrow money or ever be in debt since they can coin their own money, will start to borrow money anyways from the bankers and traders and then the governments find themselves in debt to these various <u>Order-People</u>, who now start to control even the governments and the various nations themselves. Again, since governments always have the power to create and control their own money supply, there is never any reason for a government to ever be in debt or to be forced to borrow money, especially if they can simply coin it. However, the Order-People do not want this to happen and so once a government does that, it must be stopped at once. Two current examples in the country of America are that of President Abraham Lincoln who printed Green-backs during the American Civil War and President John F. Kennedy, who printed U.S. Government Notes, both free of debt and free of control by the traders and the bankers. Both examples ended terribly with both leaders being assassinated and removed from the scene, so the traders and bankers could regain control and stop this process through their successors.

Lastly, once the governments are under control of the Order-People and the nation is in debt with huge taxes, then massive trade agreements will be founded which only allows the Order-People to take more and more wealth and control more and more things. Today, trade agreements are called Free-Trade agreements, even though they are never free and usually hurt both countries that are being forced into them. Since international agreements always overrule any individual countries own rules and laws, then each country will lose more and more of their sovereignty with each agreement.

It is at this point, that the very reason that a government exists in the first place, which should be helping the people to be and stay self-sufficient, will break down and stopped happening. Ask yourself - if your government really cares about their people then wouldn't they insure that each person and family get and then maintain a home or shelter and food or water, which again is the three things every person needs to survive? If the government you live under, taxes the very home or property that you live in and also prevents you from gardening or growing your own food or prevents you from drilling your own water well, then ask yourself whether your government even cares about its people or have they sold the nation out to the <u>Order-People</u>, and are working instead for power and control.

<u>Lawful Money as a commodity will be removed</u>

As any nation or country moves from <u>Chaos</u> to <u>Order</u>, away from visible physical things to invisible information and ideas, the money systems will also change from visible real commodities to invisible ideas.

The overall goal of <u>Order-People</u> is to eliminate the <u>Chaos</u> from society that is unpredictable and chaotic.

They are not bad people, but they truly believe that they are helping society through their actions. All physical things as a part of Chaos cannot be controlled by definition, so eventually money that is physical coined commodities must also eventually be eliminated.

How physical money is eliminated and replaced with invisible make-believe money starts first with the physical money. As described earlier, this physical money is at first made with a commodity such as gold, silver or copper and it will be coined, weighed, measured, sized and then stamped to read measured by the country to be exactly one ounce of 0.99 silver for example. This branding of the physical Lawful Money will suddenly be eliminated by changing it to only stamping of physical Lawful Money, whereas the words will be changed to read "owned by the country", "property of the country" or simply the name of the country will be stamped on the coin. Note that at this time, the word Coin will also switch from a verb to a noun at the same time.

When this happens, the physical commodity can no longer be owned or controlled by the people, but instead the commodity becomes the property of the government with which the people then only have an interest in using the commodity that the government owns. Eventually, the government through the controlling <u>Order-People</u> will start printing paper certificates that can be exchanged for the commodity, such as gold certificates or silver certificates, whereas the piece of paper can be exchanged for the commodity gold or silver. This certificate will start to be used by the people as it is more convenient to be used instead of coins.

Then over time, the government will remove the commodity backing from the certificates so that they can no longer be exchanged for the commodity. Then the <u>Lawful Money</u> will also cease to exist because it will then be converted to Debt-based money, called a Note or an I.O.U. which is a debt obligation of the government and therefore the people. The <u>Lawful Money</u> that originally had the two pillars of reality, actual physical value and actual physical form of money, will be reduced to <u>Legal Tender</u> pieces of paper that only have the form of money but not any actual real value.

It is also at this point that the <u>Legal Tender</u> money will be forced upon the people by law. This has to happen because in the real world, nobody would use debt as money since it has no inherent value and in reality, is negative money. Nobody would use it unless, of course, the government forced you to, so laws must be passed forcing the people and businesses to accept the debt or paper notes in exchange for goods or services.

Lastly, the Order-People will slowly or through the force of law, take away the right of governments to issue their own money. They do this by setting up Central Banks that are not controlled by the governments or the people, but instead are private corporations not owned by the governments or the citizens. These private Central Banks will have names that should like they are part of the government, but they are not usually part of the government and these private banks will then start charging interest which the countries must now pay on the debt-based money that the private banks will issue for that country.

One current example of Lawful Money changing to Legal Tender

One recent example of a Chaos-Based society being converted into an Order-based society that started out with real commodity based Lawful Money but that ended up with fake Legal Tender based money issued through private

Central Banks, would be the UNITED STATES OF AMERICA, as a country and as a corporation.

A few decades ago, Lawful Money was used in that country in the form of silver and gold coins. Then silver and gold paper certificates were issued by the government, so that the people used and that could be converted to the actual silver or gold commodities.

The Order-People slowly converted the banking system from one where the government issued constitutional real money to one where a private Central Bank took over the issuance of paper money. This private Central Bank also created booms and busts in the economy on purpose because once you control the money supply you can create booms and busts on purpose by limiting or expanding the money supply. These created economic problems, such as the Great Depression, which then forced the government outlaw the possession of physical gold (under President Roosevelt) which also eliminated the silver certificates and the gold-certificates.

Today the UNITED STATES OF AMERICA as both a country and a corporation, no longer uses Lawful Money, but instead uses Legal Tender called Federal Reserve

Notes, which are a debt-based currency that are issued when the private bank called the Federal Reserve tells the Treasury Department to issue them.

The actual government no longer has the right to issue their own money, so they are deeply in debt and pay the private Central Bank interest every year because they are charged interest on the debt-based currency that is issued. Since the currency is Federal Reserve Notes, as debt-based money, the people of the country and even the country itself can no longer own anything because they cannot pay for anything, because there is no real money that has actual value.

For more information regarding this last example, please consult the book, "They Own It All (Including You!) by Means of Toxic Currency". (MacDonald, Ronald; Rowen, Robert M.D., C.E. 2009)

Balance of Chaos and Order within Money

Throughout history, any country that helped its people succeed the most and helped its people stay free are the ones that used real Lawful Money or one that was backed by a real commodity that had value. Usually the monetary standard was based on gold or silver, mainly because they were the scarce commodities that are also difficult to manipulate or fake.

Keep in mind however, that there is a huge downside to only using <u>Lawful Money</u> that is tied to a commodity and that is that any expansion of any country will only be slow and honest. It is impossible to speed up and grow a country at a superfast rate, unless the scarce commodity suddenly became plentiful. But really any country that cares about its economy should only want slow and steady upward growth, because that same country that also experiences a downturn in its economy will also experience a downturn that will be slow and honest, instead of at a superfast rate.

The truth is that the actual money of any country should be physical and real as a part of <u>Chaos</u>, but the creation of this actual physical money should also be balanced with the government's ability to create digital money, without debt, at the same time as a part of <u>Order</u>. These two types of money can be balanced against each other in some way to prevent the actual amount of money outstanding from growing too fast and creating a huge inflation problem. By balancing the two types of money that must always be controlled by the government and not ever privatized by outside sources, then the Eternal Conflict can hopefully be kept in check and the country can start to succeed.

Lastly, the government must also own the banks and no bank or financial business should ever be allowed to grow or expand to a company that is larger than 2 or 3% of the entire GDP of the nation. That way the Order side of everything within the country will also not grow or expand too large and offset the balance.

With all this in mind, the next chapter will discuss language and the way language develops under the force of Order or the force of Chaos.

Chapter 8 - DIFFERENT LANGUAGES AND KNOWLEDGE

Different Languages based on Chaos and Order

All creatures with intelligence have a need and desire to convey thoughts and ideas to others. The method that we as Humans use to communicate is called language and it is a mixture of verbal sounds and non-verbal expressions, such as hand signals or facial expressions, plus written words or symbols.

Most ancient societies usually develop a verbal language as a means of communication and then eventually develop a written language to help convey the spoken language into written documents. Written words speed up the process of education, learning and knowledge, because it is much easier to refer to printed scrolls or books, especially when you are trying to teach small groups. Also, multiple copies of the same book, scroll or written word can be read over and over by more and more people and so that knowledge can spread faster.

Simple societies spoke words and told stories and the storytellers usually became the entertainment for the group or society. Verbal histories of certain societies were sometimes handed down from generation to generation by the storytellers until a written language was developed and then histories could be recorded by being written down.

The types and styles of written languages that have been developed by the different societies once again have been created based on the Eternal Conflict with Chaos-People first developing Hieroglyphical Languages were one sign, image or symbol were used to represent one specific thing. Hieroglyphical languages would literally have thousands and thousands of pictures or hieroglyphs each representing one word, item or object. Hieroglyphical Languages may seem confusing or chaotic, but they worked very well for the type of Chaos-People that used them. There are also Logographical Languages, where a glyph or logogram represents a word or sound and although these are usually Chaos-based languages, they can be part of a shift or change in a language from Hieroglyphical to Alphabetical, which is a change from Chaos-based to Order-based languages.

Order-People usually always develop Alphabetical Languages whereas there would be a total of 20 to 35 letters or marks and each one of them would represent a simple verbal sound. These languages are very controlled and organized with very little deviation in usage.

There is also a third type of Language that is called <u>Phonetic Language</u>, which is also a mixture of both <u>Order</u> and <u>Chaos</u> and is where there are both letters, marks or symbols and there are usually combined based on more complex longer sounds or syllables and these would have a larger set of between 80 to 100 letters, marks or symbols.

True <u>Hieroglyphical Languages</u>, such as the <u>Ancient Egyptian Hieroglyphs</u> or the <u>Ancient Mexican Mayan Symbolic Language</u> were used to help preserve and protect the free chaotic nature of humanity whereas most of the <u>Alphabetical Languages</u>, such as the <u>Latin Language</u> within the Ancient Roman Empire where developed to document laws and control people. In the next chapter, you will also discover that <u>Alphabetical Languages</u> are almost always used today, with their upper and lowercase letters to trick people into given up freedoms and rights.

Throughout our world's history, you will also discover that <u>Order-People</u> tend to view language and knowledge through a restricted view of control, so they will try to limit speech and censor books or written documents. They also like to restrict teaching and prevent certain things from being taught. They are very control oriented and believe that it is in the best interest of the children or students to be limited.

Chaos-People tend to like very open education that allows for open discussion and free speech. They understand the chaotic discussion can often lead to major disagreement, but that is okay for them, since disagreement is welcome even if they are the ones that disagree with the content of the conversation. That is the very nature of free speech, which is acceptance of any topic, even if it is disagreed with.

Knowledge as it moves from Chaos to Order

As mentioned earlier, the ancient story in the Holy Bible of the Garden of Eden is found in the first book called Genesis, where the Garden itself was an area of Chaos, a physical garden of physical plants, physical trees and physical animals. To balance out this garden, there was one huge tree call the Tree of Knowledge which contained everything based on Order, hence its name which included the knowledge of everything. The story also has a __God of Order__ that knows everything but wants to keep the information controlled and away from the people.

This __God of Order__, represented by the wind or the air, even threatens the inhabitants of the Garden that they will die if they take any of the fruit of the __Tree of Knowledge__ because to this God, knowledge in general has been made off-limits. The story also has a __God of Chaos__, represented by a physical snake, who supposedly tricks the inhabitants of the Garden into taking and eating the fruit from the __Tree of Knowledge__ and therefore learning things. Once that happens, the inhabitants are forced to leave the __Chaos-based__ garden and move out into the world as the knowledge that they have gained moves them toward __Order__ and away from __Chaos__. One of the many morals within this story is that as information increases in any society then that society moves away from __Chaos__ or the physical world and toward __Order__ or the world of ideas and thoughts and this includes language and words.

Throughout the world today, there are many different languages that have all derived from either __Chaos__ as __Hieroglyphical__ or __Logographical Languages__ while __Order__ usually derives __Alphabetical Languages__. As commented on earlier, some early cultures that started with a **Hieroglyphical Language** would eventually change to a **Logographical**, then later to an **Alphabetical Language** once the society moved from a **Chaos-based** society centered round hunting, gathering or agriculture to an **Order-based** society centered round construction, industry or information.

All words have meanings but get confused

Ideas and thoughts that form within the human brain use **Ethereal Energy** like electricity and therefore are part of the force of **Order**. However, even if the idea is perfect and ordered within the human brain, the meaning of the thought or idea can quickly and easily get confused, especially when the physical mouth that is attached to the physical body tries to put the thought to words or communicate the idea. This has happened to many people and is a classic example of **Order** against **Chaos**, attracting but conflicting.

Remember that it is the **Chaos** portion of the body, the human mouth that physically tries to send the audio words and language being spoken, so the words when spoken will usually be distorted, plus they can also be distorted when they are received by the physical human ear. In addition to these, the audio portion of spoken word will enter the environment and will travel as sound waves, through both the physical air made of <u>Chaos</u> but also the <u>Ethereal Energy</u> pressing down around all of us.

If you simply change the tone of your voice to make it loud or deep or use a lot of negative words instead of positive words, then you can change the outcome of a conversation or discussion. Plus, lots of people use sarcasm, exaggeration or lies, so just because someone says it or if it is written down, does not make it true. Even ancient manuscripts and ancient writing can be completely wrong, just because they are old does not mean that the writer was telling the truth at all.

<u>Agreements and the right to contract</u>

As we discussed earlier, early <u>Chaos-based</u> societies usually did not use money but instead used a barter system of exchanging real physical products for other real physical products. When the barter system was used, for example to exchange 2 fish for 1 chicken, then both products had to be real, they both had to have value and the exchange had to be agreed upon by both parties. There was usually nothing written down with this agreement, so it was called a verbal agreement, but it still represents each of the parties right to contract and to make agreements. This may seem strange, but we as humans use this right to contract verbally and to make agreements - literally all day long, every single day. Every time we pump gas into our car and then pay for it later, or every time we sit down at a restaurant, eat the food and then pay for it after it has been consumed, we have made a verbal contract with the other party that we will pay for the product or service later, even though the product was being made available immediately. Again, nothing is written down, but most societies would not survive without the individual's right to contract or not to contract, which means everyone should have a right to say "no" and to not contract or agree.

Often the right to contract can become complicated and so the agreement will need to be written down and signed by both parties, indicating that both parties agree to the written specifications. Usually these agreements are made by <u>Order-People</u> and are used to control the outcome but mostly to prevent confusion between the parties because written agreements are usually longer and more complex and may extend for a significant amount of time. Remember that verbal language can be confusing and difficult to understand, so any verbal agreement can become confusing for either party. There are some cultures today that understand the power of words and language and will refuse to contract or agree with almost anything, mainly because they are afraid to fail or not be able to hold up their end of the agreement.

Most <u>Chaos-People</u> rely on verbal agreements while <u>Order-People</u> want written agreements and contracts. Once a <u>Chaos-based</u> society moves toward an <u>Order-based</u> society, the more written agreements, laws and regulations will be compiled and put in place. Written agreements or contracts require even the language and more specifically individual words to be defined, so there is no confusion within the agreement. It has gotten to the point with most <u>Order-based</u> societies, however that there are now multiple dictionaries, one for regular language and at least one more for legalize language, which are the words of law. In the UNITED STATES OF AMERICA today, we have a Webster's dictionary for regular common language and Black's Law dictionary for legal language used in courtrooms. Keep in mind that any society does not need multiple dictionaries. They are just not necessary, unless you want to confuse the language of the people as just another means to control. When a normal person enters a courtroom in America today, they think that they hear normal language, but every word spoken, within the context of the courtroom, actually has a completely different meaning. (Webster, Noah, C.E. 1806) (Black's Law Dictionary, 10th edition, C.E. 2014 - originally composed C. E. 1860)

Censorship of words and language

As we mentioned before, words are language and they convey meaning, instruction, education, agreement or disagreement. It is a great thing to have a society that allows language to be used freely and without censorship, but only if the words being spoken do not incite violence or cause harm or damage to other people or things. Chaos-People usually understand that and allow people to disagree and state their disagreement as they believe in freedoms, rights and liberties, even of people that they disagree with, while Order-People will try to censor some language, even making some spoken words against the law to even speak.

If you live in a society that bans certain words, certain books, certain ideas, then you live in a society run by Order-People that censor these things not out of hatred or spite, but with a true belief that certain words and phrases are wrong and should not be allowed.

The main reason for censorship is that some people that have lots of Order within them, believe in the incorrect idea that all people should be happy, and all things should be fair. They truly believe that if a word or phrase upsets somebody somewhere, then it must be banned as that is not fair to that one person or one group. The reality however, is that half of everything is Chaos and so nothing is fair, and nobody can force other people to be happy. A rich person will always get a better access to the judicial system or the best college or a better house, etc.

I believe that we should not allow censorship at all or at a minimum extremely limit any form of censorship at all, because freedom, rights and liberties are important to allow all people to succeed and censorship interferes with this. Also, there are certain subjects, certain books, certain movies, certain songs, certain drugs or certain ideas that have a balance of Order or energy with Chaos or matter and those items affect all people in a positive or a negative way. but these help to open people's minds and help societies balance and grow and prosper. We will go into more detail in just a minute, but all the items that are completely balanced are usually the first ones that people want to be censored or banned.

Language as it is affected by Chaos or Order

Both sides of the <u>Eternal Conflict</u> affect each other in every aspect and that definitely include language and words. When you think and ponder about a subject that is a part of the <u>Ethereal Energy</u> or force of <u>Order</u>, by studying concept or the idea of something, for example, then the <u>Order</u> process of thinking or studying will literally change your physical surrounding and sometimes even you physically as well as mentally. The more you read and study, then the more you become open-minded to learn and change. There have been studies that have proven, for example, that prayer works in very large groups and physical people have been healed or changed by the prayer that happened or within the environment that is being concentrated on. I believe that this is not a Deity or God intervening in everyday life, but the causal effect of <u>Ethereal Energy</u> from the minds of the group of people affecting the physical environment.

It is the main reason, that open information and uncontrolled positive thoughts are wonderful and the process of keeping ideas and information available to anyone creates societies that are incredible, wonderful and innovative. When information is not controlled then all the people will benefit and grow and the society will prosper, but it is very difficult to happen, since information societies are usually all <u>Order</u> and they usually want and desire to restrict information which is one of the ways that control is obtained, but if governments and groups allowed the free flow of information then there would be less control, but the society would greatly benefit.

Regarding governments, whether based on <u>Chaos</u> or <u>Order</u>, the government should share information and legislation ahead of time and let the people read everything before new laws are made as any law should be a slow deliberate process anyways. Also, the government must make sure that news organizations and the media must not be controlled by a small group of people or the government and any one news organization must never be larger than one or two percent of the Gross Domestic Product of any nation. That way there is much less misinformation or disinformation.

Educational Systems and the Eternal Conflict

As you can see, words and language are governed in many ways by the Eternal Conflict and its relationship between Chaos and Order. However, there is one last section about this issue that needs to be discussed and that is education and the control of education.

If you were to go back in time, to some of the greatest Republics that every existed, such as the Ancient Greek Republic or the later Roman Republic, you would find that under those Chaos-based styles of government, there was no formal educational schools or colleges. This may seem strange to people today that are used to government-run schools with government approved lesson plans and homework where education and knowledge are taught to children by total strangers, called teachers, that have been given temporary authority over them during the school hours.

It may also seem strange to note that today's word School is derived from the ancient Greek work Skole, but unlike the modern word, Skole in the ancient Greek world was defined as leisure or relaxation. To the ancient Greek Republic, to learn meant simply to spend time in a garden, desert or a beach just thinking and reflecting on life, on God or on nature and then discussing your thoughts with other people. One of the greatest Greek philosophers, Plato knew that the best education came from just thinking and listening to others and so his specific <u>Skole</u> was just a place to hang out, talk, discuss or just sit and contemplate about life.

This idea of the best school being one where a student just hangs out and thinks for themselves is one that complements the <u>Eternal Conflict</u>. Just sitting around in the <u>Chaos</u> that is the physical world and using the <u>Order</u> or <u>Ethereal Energy</u> within the human brain to ponder and think is the best of both sides and as happens with all other systems, great education happens when both sides of the <u>Eternal Conflict</u> come into balance.

Remember that we also discussed earlier how there are two sides to the human brain, the chaotic <u>Id</u> and the logical order driven <u>Ego</u> and their relationship called <u>Super-Ego</u>. What we mentioned but didn't discuss in great detail, is that both sides of the human brain communicate with each other. In fact, both sides of the human brain talk and hold entire conversations with each other. Stop for a second, be really quiet and just listen to your brain right now and you will see that there are two sides of our brain and they are discussing things and holding conversations right now.

Once you understand that your brain can go into balance and when it does balance itself out, with equal amounts of Chaos and Order, you will then imagine amazing things, but only when that happens.

This is also one of the reasons that Order-People teach us that if you hear voices inside your head, then you are crazy or insane, whereas the truth is that you have always had two voices inside discussing everything. These two voices or two sides of the Eternal Conflict that are inside your very brain are the secret to learning anything and everything, because knowledge and information is part of the Ethereal Energy systems within the universe and all that electrical knowledge, of the past, future and present, everywhere can be tapped into and accessed - but only if you are in balance and communicating freely.

If you don't believe this, then ask yourself the question of where how an Idiot Savant can instantly play music or recite poetry or do extreme mathematical calculations without any training. I would propose that they are born with a specific partial mental balance, in a certain section of their human brain that allows them to effortlessly access the electrical <u>Ethereal Energy</u> force called <u>Order</u>, which also contains the vast storage of all information, past, present and future.

    This is also one of the reasons that the governments today, which are mostly <u>Order-based</u>, do not want an educated population of citizens that can perform critical thinking and will analyze what everyone does, including the government. Today's government-run schools slowly condition you to stop being curious, to stop questioning and to stop communicating from within. This is done by not allowing self-thought and internal communication as the ancient Greek and Roman Republics did, but instead they have you to going to strange buildings, with strangers called teachers that make you dependent upon them as the authority in charge of knowledge and language. The educational systems of today are all <u>Order</u> systems that teach you to seek answers outside of yourself and to blindly believe that all the authority figures put around you have the knowledge and the answers, although in reality they don't even come close.

Much like the two types of religion, there is an educational system set up around <u>Order</u> that involves belief, but balanced belief using both sides, which is when people make decisions internally backed by some outward facts, then both of these are argued and debated internally by both the chaotic <u>Id</u> and logical <u>Ego</u> of the human brain, to test the ideas and integrity of the belief that you are developing. Belief systems requires facts and effort but helps to directly develop a person's self-esteem and self-confidence, which are both important to critical thinking.

The other system, based on <u>Chaos</u>, is the one used today because it is the most convenient and easiest and is based on faith only. A faith system requires zero facts and zero effort and demands reliance on others externally for information and knowledge, to just have faith in other authority figures and not question them. This system over time quiets the mind and stops the internal communication that happens when both <u>Chaos</u> and <u>Order</u> balance within the brain.

Lastly, we mentioned earlier two special individuals that could control the internal communication within the brain between Chaos and Order and had the ability to access all the electrical knowledge within the universe and they are the American Edgar Cayce and the scientist Nikola Tesla, both of whom could balance the Eternal Conflict within their brain and perform amazing things. You should study them for more information regarding the Eternal Conflict.

In the next chapter, we shall discuss the most complex idea regarding the force of Chaos and the force of Order within any society and that is the idea of law and the different types of laws based on the Eternal Conflict.

Chapter 9 - THE TWO SYSTEMS OF LAW

The Tower of Babel and the Law

There is an ancient story found in the Holy Bible about the people of the world that wanted to be like the Gods, so they build a huge physical tower, call the <u>Tower of Babel</u> in order to reach the Gods, up in the sky and to be like them. The tower was build higher and taller than all the other structures so that the people could reach up into the realm of the Gods and challenge them. However, when the Gods saw that the tower was being built to challenge them, the Gods then destroyed the tower to prevent the people from being like the Gods. (The Holy Bible, King James Version, Genesis 11).

This is a strange story unless you focus on what then happened to the people at the end of the story, where the Gods also confused the language of the people to prevent them from understanding each other and thus prevented them from ever working together again to challenge the Gods.

What we have here within the ancient story of the <u>Tower of Babel</u>, is truly a story about <u>Chaos</u> against

<u>Order</u>, which is similar to the Garden of Eden story or the thousands of other ancient texts that discuss physical things against air things, <u>Land-Gods</u> or <u>Water-Gods</u> that are against <u>Air-Gods</u> or <u>Storm-Gods</u>, or visible things against invisible things, which also are all <u>Chaos</u> against <u>Order</u>. In this particular story of the Tower of Babel, physical people of <u>Chaos</u> challenge the Gods of <u>Order</u>, who are in charge from the air or the sky. They exist somewhere else and the physical people are upset about something, upset enough that they want to challenge these rulers of the air and sky. So, the <u>Chaos-People</u> build a physical tower to reach the rulers only to have their efforts stopped in the process. Again, the most important element of this story is the fact that the Gods of <u>Order</u>, who are in charge, wanted to prevent any future challenges to their authority. Since the people already know their language and words, it would be impossible to make them forget that and speak differently. So, what actually happened and that was eluded to in the previous chapter on language, is that the Gods just created another new dictionary where the same words have different meanings. They speak with the same words and language, but each word has a different meaning and so the words themselves become confused. This by itself would confuse the people and prevent them from moving

forward with any lawful or legal challenge against the authority in the future.

You may disagree with the above analogy but if you think about the court system and judicial systems of today, the average person must hire a lawyer, also called a translator, to even try and understanding the court proceedings and the **Legalize Language** that is spoken there. What is said in a courtroom might sound like common everyday words, but they are not. The court uses a completely different language called **Legalize** that has different meanings for the same sounding words and in the country of America, the judge and lawyers all use a different dictionary called **Black's Law Dictionary** that literally changes the meanings of the common words, so that the average person can no longer understand them, so that they are not able to challenge the new system. (Black's Law Dictionary, 10th edition, C.E. 2014 - originally composed C. E. 1860)

As we also mentioned earlier, there is no need for any society to have another dictionary of the same words, unless this second dictionary is just made-up and invisible, setup by **Order-People** to redefine words and phrases that the masses of ignorant people will never understand. In fact, most people do not even know that there is a second language or a second set of words that are used in the courtroom today. These same people are also unaware that there are actually two different systems of law that have been setup in the world. There is the <u>Lawful system of law</u>, setup by <u>Chaos-People</u> and the <u>Legal system of Law</u>, setup by <u>Order-People</u>. Both systems operate differently and so any person that is charged with a crime must first understand whether they are charged with doing something <u>Unlawful</u> or doing something <u>Illegal</u>, as they are two completely different things.

<u>Lawful Laws and Legal Laws</u>

As we discussed very early in this book, there are standards and unwritten rules in every society. These are usually called the customs of the society and govern for example, how you say hello and then shake the other person's right hand or whether you bend at the waist and bow toward them. Customs are usually not written down as a formal law but instead are just taught as an informal custom or unwritten rule.

However, once a society grows and becomes larger and especially more <u>Order-based</u>, then they will also start writing laws that can be enforced by some form of government official, such as a police officer. Much like all other systems in life, there are two types and the first is created by <u>Chaos-People</u> and are called <u>Lawful Laws</u>.

<u>Lawful Laws</u> are created under <u>Chaos</u> governmental structures like Republics and they are written laws, based on only physical things. They are normally written to only protect the rights of the people in the society and to punish those people that damage or harm other people or actual physical things.

<u>Lawful Laws</u> are usually based on a Constitution that is written by the <u>Chaos-based</u> people to control and limit the government, as <u>Chaos-People</u> believe that the people have inherent rights that have been given to them by their creator or some form of God. By defining the rights people as given to them by a form of God, then no government can remove these freedoms, rights and liberties, at least if you believe in a God. It is also the reason that <u>Order-People</u>, once they take power, go out of their way to try and remove God from all aspects

of the government and if they are successful then they can remove the inherent rights of the people. <u>Order-People</u> don't really care about your silly rituals and beliefs, but they care if those same beliefs affect their ability to control.

<u>Legal Laws</u> usually come later, when a <u>Chaos-based</u> government starts to become an <u>Order-based</u> government. Remember that <u>Order</u> is the invisible energy systems, so all <u>Legal Laws</u> are not based on reality, real things or even real people, but instead are based on make-believe rules around made-up entities. Speed limits for automobiles, for example are a form of <u>Legal Law</u>. Why can you get a ticket for driving 30 miles per hour on a road with a sign that says maximum speed limit is 25 miles per hour? The answer is that somebody just made up that number that is posted as the speed limit and they can change the speed limit tomorrow to a different number if they want. <u>Legal Laws</u> are NOT based on damage or harm to real people or real things, they are based on commerce and are usually defined by a fake money system that is based on <u>Legal Tender</u> in a society that has moved away from real <u>Lawful Money</u>.

Invisible Corporations are everywhere

For society to move from <u>Lawful Laws</u> that govern reality over to <u>Legal Laws</u> that govern make-believe and made up concepts and ideas, a system of make-believe corporations must first be in place for the system to work.

Today, everybody has heard of corporations and most companies that exist today were formed as a L.L.C. or Limited Liability Corporation or as simply a Corporation. The reason is that if a person takes their real wealth and starts a real company, they can be liable for all damages, lawsuits or problems that arise.

<u>Chaos-based</u> real companies have unlimited potential but also have unlimited liability and worse yet, they are subject to the <u>Chaos</u> that exists in the physical world. However, <u>Order-based</u> companies, as corporations, still have unlimited potential but have very limited liability if things go wrong and therefore <u>Order-People</u> love corporations so much. They have all the upside but hardly any of the downside.

So, what is a corporation? A corporation is a make-believe entity. Once a person or a group of people get together and form a corporation, they basically make up a name then they file a form that says that the make-believe name exists in invisible form and then they file another piece of paper attaching real factories or offices to this make-believe invisible entity called a corporation. We know that it is make-believe, because it was created out of thin air and tomorrow, the same group of people can dissolve the corporation or make it go away - poof.

All corporations that use <u>Alphabetical Languages</u> must spell their make-believe name in all capital letters as opposed to using both upper and lowercase letters. Any name spelled in all capital letters might indicate a make-believe entity or corporation. This is one of the many reasons that <u>Order-People</u> want to use only <u>Alphabetical Languages</u> and will try and eliminate <u>Hieroglyphical</u> or <u>Logographical Languages</u> from society, if they are given the chance.

What you may not be aware of, is that if you live in an <u>Order-based</u> society with both <u>Legal Laws</u> and a <u>Legal Tender</u> money system, then you as the real person also have a make-believe corporation attached to you, that you may not be even aware of. If you look at your birth certificate, driver's license, pay-stub, bank account or for that matter any legal document, you will see that your name is spelled in all capital letters. Your name spelled in all capital letters is not you the real physical person, but instead it is you the invisible corporation that was attached to you, usually at your birth when your parents submitted legal documents on your behalf, without your parents even knowing what they were really doing. An invisible corporation that is attached to you must occur in any society that has <u>Legal Laws</u>, because since the <u>Legal Laws</u> do not affect anything real, when you get into trouble with the <u>Legal system</u>, it is actually your invisible corporation attached to you that is actually in trouble, not the real you that only <u>Lawful Laws</u> affect.

This is also the reason that people get into trouble with the Legal system and then complain that their Lawful rights were violated, not understanding that there are two systems of Law based on the <u>Eternal Conflict</u>, whereas the <u>Order-based Legal Laws</u> punishes your invisible corporation attached to you and that have no rights but only governmental privileges, and the

Chaos-based Lawful Laws that will punish the real you but only if you hurt or damage other physical people or other physical things. You have freedoms, rights and liberties under the Lawful system, but only if you are not an atheist and believe in a God or Deity that gave you those inherent rights.

Laws of the Land and the Laws of the High Seas

Lawful Laws are sometimes called Common Law or the Laws of the Land, because they are based on real physical things but are only in effect within a certain geographical land area or nation. They are literally the Laws of only that specific land.

Legal Laws are sometimes called the Law of the High Seas, because they were created by Order-People forcing international trade upon different nations, where the trade goods usually moved over the high seas and oceans, usually on boats or ships. Since different nations have different Lawful Laws regarding physical people and physical items, a set of imaginary laws were created that governed commerce and trade and these set of laws were ones that each different nation, involved in trade, would have to accept. It is also why Legal Laws are also called the <u>Laws of the Water</u>, the <u>Laws of Commerce</u> or the <u>Laws of International Trade</u>.

Since <u>Legal Laws</u> are the made-up laws and make-believe ideas and rules, then much like your attached invisible corporation that you may not have even known existed, every country, every city and every state also has to become an invisible corporation, in order to be involved in trade. Also, whenever a country switches over to make-believe money called <u>Legal Tender</u>, that same country must start using <u>Legal laws</u>, because <u>Legal Tender</u> is usually debt-base money and no real person can have debt. But invisible make-believe entities, called corporations can have debt, so people must have a corporation attached to them to allow them to use the debt-based notes or invisible money. Today, you will find that when the name of a country or state is spelled with all capital letters, then they have been incorporated and have an invisible entity attached to them, governed by the <u>Legal system of commerce</u>. Also, most cities or towns have also incorporated to call themselves municipalities. So, everything around you may look real but is actually make-believe places, make-believe money or make-believe people that are in reality just make-up corporations.

Three Maxims of Legal Law

Remember that <u>Order-People</u> want control, since they truly believe that chaotic driven people need to be controlled. So, they slowly transition <u>Lawful Money</u> that is real with value into <u>Legal Tender</u> that is negative debt-based money. Then they secretly attach invisible corporations to your real self. They then create a second Legal dictionary that defines common words differently, for example the Legal dictionary defines a "Person" as an invisible corporation, so when the Legal court system in America asks if you are a PERSON and you reply yes, then you are actually agreeing that you are a corporation and not a real physical entity.

Next the <u>Legal Law</u> systems will slowly eliminate the <u>Lawful Law</u> systems to the point that Common Law or Lawful courts no longer exist or are almost impossible to find. All the paperwork issued by the Legal courtroom will automatically use your Legal name in all capital letters and not your Lawful or family name. In fact, most of the computers that are used today in government buildings, banks or courtrooms cannot even type a lowercase letter.

After all that is in place, then the court system will automatically assume the three **Maxims of Legal Law**. The first maxim of Law is **Let him that be deceived by deceived**, or in other words if you do not understand the difference between the Lawful and Legal systems or that the court uses a different dictionary, then that is your problem not the court's problem. The second maxim is the **Assumption of volunteering**, whereas if you show up to a Legal courtroom then unless you state it, you are assumed to have volunteered to be there and that you automatically desire to contract with the court system by just showing up. The last and third maxim is the **Assumption of Legality first**. This means that unless you state differently, the **Legal system** and not the **Lawful system** is automatically in affect. So, if the judge asks for your name and you give it to them, it is assumed that you are giving your uppercased Legal name of your attached corporation and not the real family name of your real self.

Keep in mind that all governments today, even State or Federal, are also corporations under the Legal make-believe system, so you have no freedoms, rights or liberties at all unless you expressly reserve them. Once governments reach the point of Legal entities, then even the Lawful Constitution that was created to control the governments are no longer valid.

Make-believe Law vs. Real Law

<u>Legal Law</u> is a part of the force of <u>Order</u> is based on invisible, made-up and make-believe systems. When you enter a Legal courtroom, there is no truth or reality. In fact, Legal experts call the Legal courtroom a blurred courtroom, since there is no truth and facts are usually ignored there. Only in a Lawful courtroom can the truth be presented and judged, since a Lawful courtroom is based on real things and real people.

Interesting, if a Lawful or Legal issue gets up to the highest courtroom, such as the United States Supreme Court, they have two different final outcomes based on whether the argument is Lawful and real, or it is Legal and fake. The final ruling on a <u>Lawful Law</u> will be called a <u>Decision</u>, because decisions are real and absolute. However, the final ruling on a <u>Legal Law</u> will be called an <u>Opinion</u>, because you cannot decide anything based on fake or make-believe invisible ideas.

One last interesting note is that once <u>Legal Laws</u> take over governments and people and systems, then even the real objects cannot be owned from a Legal view-point. This makes common sense, because everything has been switched over to make-believe and invisible and so real objects cannot be actually owned. Also keep in mind, that make-believe persons that are Legal also cannot own real objects, so the governments, which are also corporations, develop registration systems.

Registrations systems are where the Legal government gets you to register your property with them, so that you get back a certificate of title for the car, for example, or a birth certificate when you register your children. The act of registering however is the Legal process of giving away. So, in reality you give the property to the government and the government takes possession but then gives you back a piece of paper, such as a <u>Certificate of Title</u>, that allows you to have an interest in the property, so you can sell it later. Today you never get a real title, but instead get only an interest in the property that you registered, hence the word <u>certificate</u>.

The Hierarchy of Laws

Lastly, regarding the different laws, there is a hierarchy of seven different types of law that have been in place throughout history and even exist today. It is important to understand that these seven types of law have a hierarchy, with the first law being more important and overriding all the bottom six.

The first of all laws is called <u>God's Law</u>, which is sometimes called the <u>Laws of Nature</u>. We mentioned this law briefly earlier in this book. If you believe in God and are not an atheist then your freedoms, rights and liberties exist, only because of this first type of law. This first law overrules all of the other laws is also una-lien-able (cannot have a lien imposed) and inherent, which means your rights are given by God and are automatic. Since the government never gave them to you in the first place, then no government can take them away.

The second of all laws is called the <u>Law of Conquest</u>, which really means wars and spoils of war. This law also applies when there is the need for a military court regarding war crimes or the need to rule during war time.

The third of all laws is called **International Treaties** and is the right of any country to contract with another country. It is important to note that through International Treaties, both **Lawful Laws** and **Legal Laws** can be overridden since International Treaties usually take away some of the sovereignty of each country.

The fourth of all laws is called the **Monarchy Law** which usually only applies to **Oligarchy** governments and means that whatever the King or Queen wants then the Queen or King gets.

The fifth of all laws is called **Constitutional Law**, **Common Law** or **Lawful Law** and as discussed is usually created by **Chaos-People** about real things. Usually a **Bill of Rights** is also established at the same time, but people must understand that Constitutional rights are given by the government and not given by God. If you ever must fight for your rights, then fight for your rights that come from the first of all laws, which is **God's Law** and not Constitution rights, which happen under the fifth of all laws.

The sixth of all laws is called **Statutory Law** or **Legal Law** and as discussed above is usually created by

Order-People to help to control people. It is based on fictional corporations and make-believe laws.

The last and seventh of all laws is called Equity Law and is rarely used anymore except within bankruptcy proceedings and appeals courts where a final determination is needed that is fair or as equitable as possible.

These seven types of laws have seemingly always existed but some of them are repressed or elevated, depending on whether the government and the laws within it are Order-based or Chaos-based.

Now that you understand all the ways that the force of Order or the force of Chaos responds to the culture and society around it, then you can truly understand that the Scale Chart at the beginning of this book is accurate. When the force of Order rises, then with it rises up control, rules and regulations. Governments become Democracies and Oligarchies with Legal Tender money and Legal laws based on invisible corporations and make-believe ideas. Systems become closed and regulated and energy systems also become controlled. Censorship and restrictions are imposed by those Order-based governments that use alphabetical languages and secondary dictionaries as a means of further control. Architecture and construction becomes complex with the goal throughout the society to rely on the government and not be self-sufficient. Usually all of this happens at the same time until the force of <u>Order</u> becomes too large for the society and then the pendulum swings back and <u>Chaos</u> starts rising up.

The next chapter is my final thoughts and thanks. There are three chapters after that but those are mainly a review of known history described through the lens of the <u>Eternal Conflict</u>. Those chapters are included at the end to help you understand the cycles and patterns that have existed throughout all of world history, but also to show proof that the <u>Eternal Conflict</u> is very real and present in all things.

Chapter 10 - FINAL SUMMARY & MESSAGE

Physics is not that complex

Physics should be the simple study of the natural world but in the minds of many physicists and scientists it has not remained simple, but rather it has moved into this unnecessary realm of extreme complexity. Whether this is due to the necessity of obtaining money or grants or simply the standard problem of seeing the forest through the trees, we should always remember <u>Occam's razor</u>, which I paraphrased earlier as whatever hypothesis or idea has the fewest assumptions or fewest parts should be selected. Life is actually very simple, and we should not try to impose complexity upon such an undemanding system. Again, I believe that the <u>Eternal Conflict</u> is the simplest of such ideas to understand physics, at least for now. Maybe a different, even a simplified version will occur later that will be more concise and explain life in an easier way. But to that, I will add what the Holy Bible states in Ecclesiastes 1 verse 9, which is the phrase that says, "that what has been will be again, what has been done will be done again; there is nothing new under the sun." (The Holy Bible, King James Version, Ecclesiastes).

The real proof - Chaos, Order and the worldwide pattern.

To me the real question of physics is how do we know that the <u>Eternal Conflict</u> is a real struggle between forces? To that I answer that by reviewing history, which we will do in the final three chapters and it is there, where you will see that most of the ancient and even modern cultures that started out as <u>Chaos-based</u>, build simple, pyramid-type structures, used hieroglyphically languages and made the snake their emblem, symbol or God, while at the same time, the <u>Order-based</u> cultures build complex ziggurats or complex cathedrals, while they used alphabetical languages and made the eagle their emblem, symbol or God? Not just once or twice, but always throughout the entire course of history and human events.

It is the truth that repeats and repeats - when the force of <u>Chaos</u> becomes too overwhelming with not much <u>Order</u>, or when the two forces stay out of balance for too long, soon the force of <u>Order</u> will automatically happen and will eventually exert itself, with <u>Order-People</u> rising up and more rules, regulations and laws happening. Eventually <u>Legal Laws</u> will take the place of <u>Lawful Laws</u> and <u>Lawful Money</u> will soon be replaced by <u>Legal Tender</u>.

Control of the people will also happen, gradually or even sometimes suddenly and wars may break out, because whether we want to admit it or not, humanity is part <u>Chaos</u> as well as part <u>Order</u>. If there was complete balance among all people, then there will still be 50% of the people that want no control or order at all, as they are just chaotic children and even act as children, even in adulthood. These are the same people that ask for permission, but the same people that complain about trivial things, play practical jokes and throw temper tantrums as children, especially when they don't get their way. In the minds of the <u>Order-People</u>, not all of these adults will end up in prison, but even the ones that don't will leave a wake of despair behind them, with hurt spouses, abused children, divorce, alcohol, drugs and other pain, left behind them as their calling card. <u>Pure Chaos</u> is still chaotic.

Since <u>Order-People</u> see humanity as children that need to be taken care of, they do not believe in the infinite ability of each person. They believe truly that people need to be controlled from their chaotic emotions and this is why they believe that more rules, regulations and restrictions are very important. They believe in limiting the educational systems to ones that only teach limited tasks as a way to help restrict people's crazy ideas. They believe in limited or finite energy systems that people must use less and less of. They believe in debt-structured money systems, again to help limit the destructive tendencies that people have. They believe in abortions and birth control to help limit the number of chaotic people being born. They believe in complex financial and tax systems to fool the masses into compliance. Their political systems will be Democracies that become Oligarchies. They teach leadership by wealth, deed or bloodline. They also believe that all the wealth of the world must be held by only the smartest people to limit other people's ability to use it. They will also change religion or God into a judgmental invisible force, which is how they see God, or they will try to eliminate God completely because the concept of God will be counter-productive to their goals.

But once again, the force of Order will eventually become overwhelming as the force of Chaos is restricted and over time, the people will rise up and revolt or cause a revolution or worse yet, terrorism and mass destruction will happen and will occur automatically in order to bring more Chaos back into the world to try and maintain balance.

Automatic push-back of both Chaos and Order.

Order-People usually beget Order-People, which are raised without a point of reference regarding Chaos. Also, Chaos-People give birth and raise more Chaos-People, which are also raised without a point of reference regarding Order. So, both sides of humanity literally do not understand the other side, which is where most conflict occurs. They also do not understand that there is a universal truth that both Chaos and Order must be balanced or when an out-of-balance situation happens, that the force of Order or the force of Chaos will automatically push back, usually to the extreme to try and maintain balance. This push-back can be destructive and devastating.

Keep in mind that the Eternal Conflict affects everything, and both the force of Chaos and the force of Order are just as active within societies, communities, countries and even from a planetary standpoint as they are from an individual standpoint working psychological from within the human brain. When it comes to the sociological standpoint of the Eternal Conflict, the subject can become a little negative and depressing.

As I mentioned, <u>Order-People</u> and <u>Chaos-People</u> really have no point of reference when it comes to the other side or other viewpoint. They both believe that they are correct in any argument or discussion and that is because they have an active force working on their brain helping to form their ideas and thoughts. A person focused on <u>Chaos</u> believes in physical things and would normally see abortion as a bad thing that must never happen, whereas a person focused on <u>Order</u> can see the value in abortion, especially for poorer less fortunate women. This drastic difference comes from opposite forces acting upon the brain, but the truth is that both individuals are correct from their perspective, but they literally cannot comprehend the other viewpoint, so that is why things can get destructive between the two people, because of the two opposing forces at work. Worse yet, is that both forces and both opinions are attracted to each other, since opposites do attract and that is why sometimes both forces can build quickly and cause major damage.

Interestingly, the two forces that are attracted to each other, also oppose each other and things that are in opposition tend to try and tear each other apart and you can see that whenever two people that have polar-opposite viewpoints confront each other. What happens from a sociological standpoint is that one of the forces takes over and builds up, then once it becomes overwhelming the other force sweeps in quickly to try and create equilibrium, status or balance.

The most important thing to understand is that if the out-of-balance situation within any society is extreme or when the pendulum of one of the forces reaches the farthest point up on its side, then the other force will come in with great magnitude and the tearing apart of that society will occur. In human history, evolution or sociological change is never a slow process. The two forces are at work and although the need for balance is strong, it almost never happens. So, one force builds up to an extreme point and then the other force comes in and create destruction and problems. When you review history, major events, such as mass extinctions or mass change seems to never be slow or gradual. Maybe, in B.C.E. 20,000 - there were major cities powered by nuclear power plants and technology - maybe the culture looked almost exactly like it does today, but a major destructive event occurred, and the entire world plunged back into stone age. Isn't that possible? Are we so intelligent today that we can't even fathom the ancient past having people just as smart as ourselves, who could create technology the same as we have?

I do know this - that in the early 21st century, the governments of the world are mostly <u>Order-based</u>, with <u>Legal Tender</u> currency, called U.S. Federal Reserve Notes being used as the world's reserve currency and <u>Commerce</u> and <u>Legal Laws</u> being most of the world's rules. In fact, with all the <u>Legal</u> trade agreements in place worldwide, most of the civilized world has now moved toward <u>Order</u> to the detriment of everything else. In fact, I would propose that anytime a society fully embraces electricity to the extent that it is now, powering the modern world and the high technology that comes with it (such as computers, television, the internet, cell phones and mass electronic devices) that every time that society has now become fully governed by <u>Order</u>, which is the invisible technology itself. The force of <u>Order</u> rises up as technology also rises up. I would also propose that this has happened numerous times in the past with the same potentially destructive outcome, because whenever that happens, <u>Chaos</u> has been pushed aside way too far and an extremely extreme out-of-balance situation is occurring within the <u>Eternal Conflict</u>, and it is during those times that the force of Chaos will also be reasserting itself with all of its destructive potential.

Ask yourselves - Why in the year C.E. 2015 did the citizens of Britain vote for Brexit or to leave the European Union trade agreement and its accompanying Legal status? Why in the year C.E. 2017 did Donald Trump, a reality television celebrity become President of the United States by simply telling people that he would go against the New World Order, become protectionist and more Chaotic? Around that same time, why did the terrorist group ISIS suddenly rise from nowhere to begin its massive destruction of both the societies of the Middle East and even their ancient history, because they are destroying even those nation's treasures? Why in the year C.E. 2018 is there suddenly a possibility that the European Union may fail?

I would answer all those questions with the same answer and that answer is that the world had become too focused on Order and so the force of Chaos began once again to exert itself and it did so quickly and forcefully. For those Order-People that became horrified about those events as they happened (and there was many) I would say that the coming changes back to Chaos are not done yet. The pendulum is now swinging back toward <u>Chaos</u> and away from <u>Order</u> and hopefully this time around, the worldwide changes will not be as extreme as I think they have the potential to be. I say hopefully, because in my opinion, the Earth as it currently exists would not survive another world war. I would ask you, the reader, to view all the coming world events from the perspective of the <u>Eternal Conflict</u>, because only then will you be able to truly see the need for balance between the only two forces at work in the universe, which are <u>Chaos</u> and <u>Order</u>.

<u>The Summation Quotation in retrospect</u>

Lastly, if you remember the first quotation that started this book, then you should know that I truly believe that Mr. Paramhamsa Tewari is absolutely correct with that quotation especially in that we must, as a unified world, start to understand the nature of energy and the attributes of the nonmaterial Akasha which together comprise the force of <u>Order</u> and we must also start to gain the knowledge of the single material particle which constructs the material universe or the force of <u>Chaos</u>, in order to help create the balance that is needed to achieve harmony and peace. (Tewari, Paramhamsa, C.E. 1996)

Again, there are three more chapters, which layout the proof of the Eternal Conflict within our ancient past and throughout history - but before that, I want to personally thank each of you for reading through all these thoughts and ideas that are contained within these pages. As I mentioned earlier, I am hopefully that the words that you read, simply made your soul ponder and your mind think and once again, that can never a bad thing. Also remember, the most important last thought - to not believe what others say, not even the words that you just read and that are written here. Read everything and believe very little, do your own research and come to your own conclusions. Exercise both the mind and the body, because it is the only way that you, the reader, can have a wonderful life that is always in balance. Thanks again.

Chapter 11 - HISTORY OF SUMER & THE LANGUAGE WAR

The struggle of different nations under Order or Chaos

In order for you to truly understand the <u>Eternal Conflict</u> and the two forces that are work within the universe, within every individual, within each society and even everywhere in the universe, it is interesting to look at world history and view it from a macro standpoint of the <u>Eternal Conflict</u>. Specifically, different moments in history that demonstrate that the forces of <u>Chaos</u> and <u>Order</u> are at work against each other, molding and forming the world and the history that we are aware of through their interaction.

There are three chapters total that will hopefully accomplish this, but this specific chapter is dedicated to what would be considered the beginning of ancient history in the world. We are starting with history around B.C.E. 4,000 to B.C.E 3,500 and within ancient Mesopotamia in the lands of Sumer or what we today consider the country of Iraq. Please understand that there have been many significant major events and historical cultures prior to those dates and in fact archeologists have discovered many significant ancient sites that had even show advanced people and even advanced technology that date much earlier to B.C.E. 10,000 and even before that, but the main reason that we are starting around B.C.E. 4,000 is because this is the earliest recorded history that we are currently aware of.

Throughout world history, the opposition and attraction of the force of <u>Chaos</u> and the force of <u>Order</u> can be seen as footprint across the ages. You will see that most of conflicts, battles and wars have been fought due to the forces of <u>Order</u> within one country reacting to the forces of <u>Chaos</u> within another country. You will also see that the governments, money systems, religious systems and law systems all follow the same patterns over and over again. These patterns strangely use the same symbols, architecture, buildings, systems, mascots and deities that represent either the force of <u>Chaos</u> or the force of <u>Order</u> and no matter the time frame these will stay the same.

When you try and reconstruct the history of Earth, you must also look to the ancient writings, scrolls and tablets. But as you view these, you must keep in mind that some of the writing may not be true at all but rather fiction. However, you can simply view all the writings together, again from a macro viewpoint, to see if any pattern arises.

Chaos Water-Serpents against Order Air-Gods or Sky-Gods

We shall begin our review of the Eternal Conflict within ancient history in the Middle East, within the lands of Sumer in particular, as most of the earliest writing was recorded there. But also, these lands were the first to be established under the control and rule of Order. The Ruling-authority maintained control over the people but also attacked and overthrew many of the surrounding areas that interestingly were established under Chaos. By starting in this time-period and in this area, you can see the ebb and flow of the Eternal Conflict as it moves through the ages.

In ancient Sumer, their primary Deity was called Anu and this God had two sons, one was the God of Order and his name was Enlil and his symbol was that of the invisible air or sky represented by a bird, more specifically an eagle. His second son was the God of Chaos and he was called Enki or Ea and his symbol was that of physical water represented by a snake. Enki is actually shown in many ancient Sumerian statues and stone carvings as having the symbol of two serpents circling a staff from opposite directions, which is similar to the structure of human D.N.A. and is the source of the later mythological rod of Hermes (another later trickster God of Chaos) or the Greek staff of Asclepius (which only had one snake encircling a staff) but also one of the modern symbols used for a time by the American Medical Association, U.S. Army and U.S. Navy Medical Departments.

In the ancient Sumerian writings, Enlil and Enki were brothers that were always in conflict, whereas Enlil hated humanity and wanted to put humans into slavery and control, while Enki loved humanity, describes himself as our father and wants us to have information and freedom. The brothers were so completely opposite that Enlil representing Order went to war with his brother Enki representing Chaos. This battle started in Sumer but extended outward and went worldwide. Some may say that this battle is still going on today.

Interestingly, the attributes and symbols of the God of Chaos as described in the Sumerian ancient history will not change as we move further into history, although the name Enki, as the God of Chaos will change repeatedly and will be referred to by using many different titles and different names. You will also see that force of Chaos or Physical Matter will always be symbolized as that of a serpent, snake, dragon or lizard and will represent that of the Physical Earth or the Earth-God symbolizing visible Physical Matter, which is Chaos.

You will also find that the attributes and symbols of the God of Order will also not change, although the name Enlil, as the God of Order will change repeatedly and will be referred to using different titles and names. You will also see that the force of Order or Ethereal Energy will always be symbolized as that of a thunderbolt, lightning, bird or eagle and will represent the invisible air or sky or the Sky-God or Air-God symbolizing the invisible Ethereal Energy, which is Order.

In fact, the symbols of the bird and the snake as the two opposing gods are extremely ancient but will continue through the beginning of history and throughout all the cultures and civilizations known throughout the world, up to the present day. Also, you will find throughout history, that Chaos-People who fight for freedom from too much control or corrupted authority will always use snakes as their main symbol, while Order-

People that try to impose control will always use birds or eagles as their main symbol. This will also include animal symbols on flags, banners, standards, podiums and on statues that represent the country or the group.

In fact, one of the important myths of Order-People is that of a cultural hero or deity such as an Air-God or Angel from the sky with eagle wings or with an eagle present or sitting on their shoulder. This Air-God will be shown battling or fighting a Chaos monster or Demon, usually depicted as a physical snake, serpent or a dragon. This is the most ancient depiction of the Eternal Conflict, which as you are now aware, is that of Order against Chaos.

Some of the major variations of the names of an Air-God or Storm-God that always hold a lightning bolt or an eagle, and always seem to be fighting a Earth-God, who is represented as a serpent or snake, would include the following; the Jewish God Yahweh against the Leviathan; the Canannite God Baal against Yam; the Hittite God Tarhunt against Illuyanka; the Zorastrian Oraetaona against Azi-Dahaka; the Babylonian God Marduk against the dragon Tiamat; the Christian Angel called Michael slaying the dragon; the Norse God Thor against Jormungandr; the

Vedic God Indra against Vritra; the Japanese God Susanoo against Yamata-no-Orochi; and the Greek God Zeus against the Typhon.

More examples of how the names of the Chaos-God and the Order-God change throughout history are as follows: the primary God in ancient Sumer as we already discussed was called Anu, the Chaos-God was called Enki or Ea while the Order-God was called Enlil. The primary God in the Semitic Culture was call El or Yahvé, the Chaos-God was called Baal while the Order-God was called Kothar. The primary God of the ancient Greek culture was called Kronos, the Chaos-God was called Poseidón while the Order-God was called Zeús. The primary God of the Christian culture was called Yahweh, the Chaos or Physical God was called Jesus Christ while the Order-God was called the Holy Spirit.

As we mentioned in earlier chapters, you will also see as we go through history that there has always been a trinity or relationship-of-three within the Eternal Conflict. The original trinity was based on the two forces and the relationship with themselves and each other forming three relationships. That of Chaos as a snake or Earth-God and that of Order as a bird or Air-God and their relationship to each other or in combination, forming the primary God that is both <u>Chaos</u> and <u>Order</u> together as everything.

Force of Chaos and the Force of Order and their worldly aspects
===

Lastly, as we move through history, you will find that <u>Chaos-People</u> may create laws; but only a few simple laws to govern damage or harm of physical things or people; only simple real money based on actual commodities; and only simple limited governments made to protect the people and specifically the rights of the physical people.

These same <u>Chaos-People</u> always created structures called pyramids to represent their simple chaotic but free societies. A triangle is the simplest of forms that have rigidity and strength. A square is very unstable unless you add a cross member to it from corner to corner but then in reality, you have two triangles and not a square. A pyramid, as the simplest three-dimensional shape, is used to represent rigidity and strength, while using the least amount of knowledge or engineering that could be required. <u>Chaos-People</u> literally built pyramids all over the planet Earth whenever they were in charge.

On the other hand, <u>Order-People</u> throughout history create complex laws; secondary law systems; complex dictionaries and even different dictionaries that are used as law books; complex government systems; complex money systems and they almost always create very elaborate, highly skilled, technical ziggurats or complex temple structures. Complexity is a friend to <u>Order-People</u> as it tends to bring about confusion, which helps maintain control and order.

You will also see as we move throughout history, that where there are pyramids, you will find <u>Chaos-People</u> that used <u>Hieroglyphical</u> or <u>Logographical languages</u>, smaller Republic based governments, limited number of lawful laws and real lawful money systems based on real physical commodities. Where there are complex structures and no pyramids, you will find <u>Order-People</u> with <u>Alphabetical languages</u>, a controlling Oligarchy or Dictatorship type government, with many different types of Legal laws and fake money systems based on Legal Tender and not physical commodities.

Review the __Scale Chart__ at the beginning of this book for an entire list. This chart shows what aspects that __Order__ desires or that __Chaos__ desires and you will see that they are absolute opposites and are always trying to balance themselves out.

Babylon within Ancient started under Order

As was mentioned earlier, the ancient lands of Sumer, which is modern day Iraq contain some of the earliest writings known. Interestingly, these lands are also a huge mystery since the people around B.C.E. 4,000 to 3,500 were supposedly just simple groups of farmers and nomadic herders but suddenly overnight, these groups developed some of the most complex and advanced scientific and engineering cultures of their day.

According to the ancient Sumerian writings, suddenly out of nowhere the Air-God named Enlil, introduced writing, reading, trading, money, laws, religion, art, agriculture and complex government systems. These included mathematics, science, astronomy, astrology, technology, navigation and construction. These lands were founded under __Order__ with control being the center of the region and of the many complex religions that were also started under him. Out of the entire planet Earth, suddenly the lands of Sumer, however primitive in structure, overnight became the world's only super-power at the time, having vast amounts of knowledge that no other country had. It was formed under Order with a dictator-like Air-God named Enlil as the leader.

Where did this knowledge come from? Did this Air-God named Enlil actually exist and if so, where did he come from? This is the absolute mystery of the first modern civilization that we have records of. The truth is that we don't really know if Enlil was a real person, an alien, an angel or just a fictional character in these ancient writings. But we do know from their writings that many complex laws were developed; lawyers called scribes were put in place; many complex forms of agriculture, including domesticated cereals, fruits, vegetables and even domesticated animals were introduced; hybrid and cross pollination were put in place; mathematics, astronomy and even the calendar were created; engineering and metallurgy were put into practice; and money systems were put into place and then were converted into religious centers. Transportation across land, with roads and bridges, were developed, as well as boats and astronomy for the navigation over water. All the important knowledge that nations need to survive and to subjugate other nations suddenly was given to this nation.

Within the lands of Sumer that were founded under <u>Order</u>, a large city of Babylon was created as the center of the many complex religions and money banking systems, while another large city of Nippur was created as the center of the political districts. All these cities contained huge elaborate ziggurats. None of the cities within Sumer contain pyramids or pyramidal structures and all of the people used a complex <u>Alphabetical language</u>.

Regarding the <u>Eternal Conflict</u> within the Sumerian culture, there is a lost story of the Sumerian king called Etana who was seeking the tree of life and who was carried by an eagle up to the heavens. Within the story, both <u>Order</u> and <u>Chaos</u> were in balance when the eagle and serpent lived peacefully together, but then the eagle decided to eat all the offspring of the serpent, for which the serpent rose up cast the eagle into a great pit from which it could not escape. The king Etana was the one to rescue the eagle from the pit, which symbolizes <u>Order</u> taking back control.

These ancient lands of Sumer were very instrumental in trying to exert control over its nation and then all of the Middle East lands in general. Since the lands of the Middle East unite Europe, Africa, Russia and Asia, they then sit on the very crossroads of many cultures and many lands. This position allowed Enlil and the force of Order to move over into India and into parts of Africa and Europe as they conquered other lands and expanded their control under Order. Believe it or not, this expansion of the country of Sumer to go out into the other surrounding lands is actually a normal part of Order and control.

Indus River Valley started under Chaos but became Order

At approximately the same time-period that the lands of Sumer were being established under Order, to the east of Sumer was were the western lands of India, in what is known as the Indus River Valley area, was also being established but under Chaos. These were called the Harappan culture and started out with intensive settlements along the Sarasvati River, which no longer exists today.

The urban culture of the Harappan civilization was very organized with cities setup in large grid patterns with very detailed rivers, irrigation and plumbing systems. The entire civilization was based on <u>Chaos</u>, with one of the earliest versions of a <u>Republic</u> form of government known. There were no large structures built for the government officials because the government was decentralized with power given to the people. This culture also worshipped the Nagas, which were the serpent gods and goddesses of <u>Chaos</u>.

They used a <u>Logographical language</u> that used pictographic inscriptions usually written on small soapstone squares with motifs of animals or people that meant concepts to be read. One example was a motif of a man sitting in a yogic position that meant the Hindu religion. Later these pictographs would also include letters or partial words called Indus script.

Most of the most ancient writings that were written around that area described large battles including air support, high technology and possibly even planes or spacecrafts that attacked the region because of their culture and beliefs. The most detailed of the ancient writings are found within the Hindu Sanskrit scriptures called the <u>Bhagavadgita</u>. I am not sure about many of the fantastic writings, but I do believe that the attacks came from the lands of Sumer as <u>Order</u> invaded and expanded from Sumer to the west into both Persia and India. I believe this because once the invasion was finished, both the trading and money systems were changed within this region of India into those similar to those from the regions from Mesopctamia and Sumer. I would describe this as the first example of <u>Order</u> vs. <u>Chaos</u>, Enlil fighting Enki and the invasions and battles that resulted were horrific with the force of <u>Order</u> winning.

It was also during this time that a new Hindu deity called Garuda or Syena came to prominence. Garuda was the <u>Order-God</u> and was depicted as a golden bird that had the body of a man, but red wings and an eagle's beak. Garuda was also massive and was considered the sworn enemy of the Naga serpent gods, hence Garuda's habit of feeding on snakes eternally. In the Hindu religion, an image of Garuda is believed to protect the wearer from snake attacks and snake venom. Interestingly, Garuda came into the lands of India forcefully and imposed control so strongly that Garuda is often depicted as extremely dominant and controlling. Garuda is so controlling against <u>Chaos</u>, that Garuda is often shown as wearing small serpents as toe-rings, ear-rings, belts, hair-pieces and necklaces.

During the invasion of the Order-God into the Indus River Valley from the lands of Sumer, many of the people in that areas rejected Order and the Order-based systems and traveled up and into the lands of China to escape. It was around this same time that China existed around Chaos with Chaos-based systems because of this migration. The stories of an Order-God as a massive bird called Garuda, extended up into China as people moved north from India. In the Buddhist religion and mythology, the Garuda is considered to be the ancient bird-like god that created rules, regulations and social structure. Garuda are also known within Buddhism as the enemy to the dragons and serpent people which they hunt and devour. We shall discuss ancient China later, but interestingly China as a nation existed under Chaos longer than any other nation. Therefore, there are more pyramids in the country of China than anywhere else in the world, in addition to their constant reverence and love for all snakes, serpents and dragons.

Once the Indus River Valley was forced into the trading and accepting the money systems of the Sumerians, it didn't take long for the Harappan cities to be changed dramatically. Soon these cities were rebuilt with at least one complex citadel, like the ziggurats constructed in Babylon. These citadel areas were used for the central religious area, which of course, also began to also be used as the public administrative centers for the government.

So, the Harappan culture that started off under Chaos was swept away under Order and became a commercial Oligarchy with huge amounts of commerce and trade. The new ruling elites controlled the cities and also the huge trading networks with all the surrounding areas and the Middle East.

Early Egypt was established under Chaos but became Order

As mentioned earlier, the ancient Sumerian people came from nowhere and suddenly had advanced knowledge of almost anything and everything. Then this advanced civilization, which was the Order-based country of Sumer started to battle the Chaos-based nations around it, eventually expanding eastward into India and Persia and then westward into Egypt.

The ancient Egyptian culture did get attacked by the Sumerian armies on multiple occasions throughout history, with the first being a much earlier time period, during the times when the great pyramids of Giza had already by erected under a <u>Chaos-based</u> government that also created a very detailed <u>Hieroglyphical language</u>.

As described earlier, the very technological battles of Sumer against India were very swift with the force of <u>Order</u> coming out as victorious. Unlike those battles however, the fighting against Egypt were not very successful at first. We know this because during these times, the two regions called Lower Egypt and Upper Egypt were still united and they both fought together as one country. This was the time that the country of Egypt was still <u>Chaos-based</u> under the King called Namer, who worked to help his people stay free and become self-sufficient. It was also during his reign that the symbol of the snake held great power and the construction of over eighty additional pyramids were built along the west side of the Nile Valley, even though none of those would ever match the size or scale of the original Giza Pyramids from the much earlier time.

Egypt did eventually fall to the invasions by the Sumerians and we know this because sometime around B.C.E. 2,200, the lands of what was called the Old Kingdom was suddenly divided again into two parts. The old Kings were removed, and the new leaders called Pharaohs took control of the Egyptian lands. These Pharaohs proclaimed that they had a "divine right to rule" as dictators under Order and control, because they believed that had power and authority that came directly from the Order-God. The extreme reverence that the snake had earlier in Egypt was also pushed aside and suddenly the God Horus, known as the Sky-God, took prominence all over Egypt and his symbol was that of a falcon or a bird.

Under the new Order-based Pharaohs, Egypt stopped all pyramid construction and focused instead on building very complex temples and tombs. In addition, the construction of obelisks began, which are simply a small little pyramid set on the top of a very tall square pillar, symbolizing the reach of the Air-God into the sky or air and which pushes the Chaos-God pyramids up and out of reach of the people. During the same timeframe, Egypt started using a more simplified form of Hieroglyphs, which were called Hieratic and this simplified language was immediately adopted for use in contracts and laws.

Once Order took strict hold on Egypt, the Pharaohs became absolute dictators with total authority and control and oversaw the military, treasury and court systems. The government under the Pharaohs developed huge numbers of lawyers, called scribes, to write down financial records and to coordinate the collection of taxes from the people, while working together with the high priests in the royal palaces.

Northern Peru and Mexico started under Chaos became Order

As mentioned earlier, the two forces of Chaos and Order are opposites with a mutual desire for balance, hence their attraction to each other. Since the universe strives for balance, you must understand that as the force of Order socially and politically expanded outward from Sumer and grew into what would eventually be called the ancient Assyrian and Chaldean Empires, other nations must and did automatically rise up based on Chaos, because the two forces are always at work and in a universal way, they are trying to maintain balance around our planet and across the universe.

So, during the time that the Order-based Assyrian and Chaldean Empires were expanding and reaching out of

Sumer and into Persia, Egypt and India, the Andean civilizations in South America and the Mesoamerican civilizations in Central American rose up and they were mostly all based on Chaos.

The Andean civilizations were called the Norte Chico and Caral Supe and were in northern Peru. These were the same Inca people that would later establish the Paracas culture and who would later create the mysterious Nazca Lines of Peru. These people had one primary religion which was the monotheistic belief in one true God with divine powers and true wisdom. They called their God Viracocha or Wiracocha and he was considered the Snake-God and can be described in similar water terms as Enki was. For example, this Chaos-God called Viracocha was intimately associated with waters and the sea.

The dual natured Eternal Conflict was a very prominent theme within the Inca culture and it was described as the duality of the Cosmos or the conflict between Hanan Pacha or the Heavens and the Uka Pacha of the physical universe.

It was under the direction of the Chaos-God called Viracocha that the Inca economy had vast storage systems for grain and foods and their markets, called Catus, relied only on bartering and exchange without any regulations or control. In fact, for thousands of years, the Inca culture refused all forms of trading, money or banking systems.

Similar to the early Chaos-based Harappan culture of India, that were eventually attacked and taken over by Order, the Andean civilizations of Peru have many ancient writings that also describe many large battles that were fought between Viracocha, the Chaos-God of water and Inti, was considered the God of lightning and thunder in the air, which would be the Order-God.

Viracocha was a very beloved God represented by snakes. As the social force of Chaos rose up during these time periods, the Water-God or Chaos-God, represented by a snake also became very prominent as a major Deity within all the Andean cultures. The Serpent-God was also known through the Mesoamerican cultures of Mexico and Central America, but they used a different name for him. The Mesoamericans cultures were called the Mayans and another culture called the Aztecs and both societies called their Serpent-God by the name of Quetzlcoatl and again that was just another identical version of the same <u>Chaos-God</u>.

During the reign of the water or serpent God Quetzlcoatl in Mesoamerica and the serpent God Viracocha in the Andean culture, both societies built many simple structures, including many ancient pyramids and large pyramidal structures. The reverence for the snakes was so strong within both cultures that you can even today see literally thousands of snake or serpent statues and figures within those lands.

It wasn't until later, when the <u>Air-God</u> or <u>Order-God</u> took over those lands, did extremely complex ceremonial buildings get constructed and it was within these sacred sites that suddenly many statues, pottery cups and other physical objects became a part of their new materialistic religions led by spiritual leaders. It was also during the later times of <u>Order</u> that the <u>Air-God</u> or <u>Order-God</u> would demand many blood sacrifices and child sacrifices.

<u>The Canaanites were neutral in the battle between Order & Chaos</u>

Around B.C.E. 3500, the language and trading systems setup in Sumer, under the <u>Order</u> systems of Enlil, found their way to the north into the lands of the Canaanites, which is modern day Israel, Syria and Palestine.

The lands of Canaan originally were setup under a balance between <u>Chaos</u> and <u>Order</u> that allowed them to be neutral. The lands of Canaan were mostly held by farmers or herdsman that controlled most of the lands to the east of the Mediterranean Sea. But as trading evolved into these areas, the lands of Canaan became important and valuable due to their specific location. This land happens to be in a very strategic place bordering Africa, Europe and the Middle East and although the Canaanites were mostly agricultural to begin with, their strategic location started to be used and merchant cities began to be built along the coast by the many different traders from Sumer. Due to their partnership with traders and bankers, they were one of the first <u>Chaos</u> nation to adopt an <u>Order-based</u> language, which was the <u>Alphabetical Language</u> used by the traders.

As the merchant cities started to rise up, the Canaanites were wise enough and actually started to get in on the action that was happening in their own lands. They did this by helping to facilitate trade between all the neighboring countries and more importantly, they were one of the few countries to not be attacked directly in the all the recorded battles between the Order-God from Sumer and the Chaos-God whose free societies, such as Egypt and India were being invaded and taken over.

Again, the Canaanites at that time were the first nation to not take a stand but rather stay neutral. This meant that they honored both the Order-God and the Chaos-God, but they did not worship them directly. They understand the wars and battles being found by both Gods over the different societies, religions, politics and commerce. More importantly, they realized the benefit of not taking sides while advancing in trading and commerce.

Some of the earliest Canaan mythologies openly discusses the struggle between Baal Haddad or Teshub, who was the God of the sky or Order and the opposing God of the waters or Chaos who was called Yaa, Yaw or Yahu.

Baal, the Order-God, took the symbol of the eagle and was also referred to as the storm God and is shown with a thunderbolt in one hand. Some considered him to be the Canaanite version of Enlil, the Order-God from Sumer, who had a wide influence outside of Sumer during this time and was considered a very mean God who started many wars and who tried to enslave people through trade, language and money.

Yahu, the Chaos-God, took the symbol of the snake and was considered the creator and the protector of all humanity. Some considered him to the Canaanite version of Enki or Ea, who was the brother of Enlil. The God named Yahu or Enki was also known by his earlier name of Akkadian.

The neutral lands of Canaan during this time, became a very prominent and important place within the major religions of Hebrewism, Christianity and Islam. They are all called the Abraham-ic religions as they all have the same belief that Abram or Abraham was their founding father and they also believe that their specific religion started and developed in this area, during these ancient times. Interestingly, it was the neutral position taken by the lands of Canaan, in these ancient times, that allowed all of these Abraham-ic religions to form and even today, many different versions of those same

Abraham-ic religions still exist in those lands and round the world.

Over time, the Canaanites would be influenced by the Philistine and Phoenician cultures, both of which saw the import and export of goods through the neutral lands to escape the heavy tariffs and taxes that the countries, under Enlil would charge for. Eventually, the neutral attitude of the Canaanites was challenged by the <u>Order-God</u> and he eventually invaded the lands of Canaan also. It is this later invasion of Canaan that took place during what I call the <u>Language War</u>.

The Sumerian Expansion and first major revolution against Enlil

As the Sumerians invaded and expanded into the Middle East and beyond, the lands were combined and called Mesopotamia. There were two major areas within Mesopotamia and the first we already talked about called Sumer or Shinar and it was on the Southern coast. The second area was in the central lands and it was called Assyria and had the same types of <u>Order</u> systems within banking, trade and laws. Within all the lands were many complex buildings, such as ziggurats but also many sphinxes that are large statues of creatures that were half man and half animal. Of course, being nations under Order, there were no pyramids or pyramid-type structures build within these lands.

During the start of the Language War, the Chaos-based India and Egypt had already been invaded and taken over. This was also the same time-period, when many of the Gods were depicted or drawn as half-man and half-animal.

The leader of Mesopotamia at this later time-period was called Nimrod and he was set up as a leader by Enlil. Since we are now talking about the time-period around B.C.E. 2150, it is safe to assume that Nimrod was either a follower of Enlil, a creation of Enlil or Enlil himself, since he continued Order-based control over all the lands of Mesopotamia. Nimrod was also called Nebrod or Amraphel in other various ancient books. What we do know about Nimrod was that he hated and had great contempt for one God called I-Am or All-ah who was known in the lands as the God who fought against Enlil. This hatred was so strong that Nimrod created multiple religions with multiple Gods, but all centered round the worship of Enlil under various names.

Nimrod was one of the first world leaders to take the idea of the Egyptian Pharaoh's that they had a <u>Divine Right to Rule</u>, and then expand it into the idea that certain leaders are literally ordained by their God to be in charge of a nation while everyone else is a minor creation. This idea is definitely an <u>Order-based</u> or <u>Control-based</u> idea, because this idea asserts that the leader or king is above all other authority on Earth and that they get their divine right to rule directly from God, who at that time was Enlil. This allowed Nimrod to no longer be subject to the people that he ruled, because he got his authority and right to rule directly from Enlil. Interestingly, this same idea took the normal rights, liberties and freedom of all the people and completely suppressed them because since the leader has the divine right to rule, then the people can only get their rights from their leaders or from the government.

Under Nimrod, the lands of Mesopotamia became even more of a tyranny, where he established fire worship, blood sacrifices and idolatry with statues and multiple Gods. You will discover that countries that have been under <u>Order</u> for a long period of time will eventually go to the extreme of controlling people and will eventually establish violent games or sporting events and blood sacrifices.

The Tower of Babel and the start of the Language War

It was under the reign of Nimrod that the ancient story of the Tower of Babel was written, which I believe was not just a story but an actual event that occurred. As mentioned earlier, this event happened in the city of Babel and is a very significant story about the battle between Order and Chaos and about early civilization communication and language. The story of the Tower of Babel as described in the Holy Bible was part of an earlier story, originally called Enmerkar and the Lord Aratta within the ancient Sumerian culture.

Both stories are slightly different, but both speak to the desire of Chaos-people who want freedom to speak one language, so that they may be understood by each other and therefore can fight against Order and control. As we discussed earlier, this Sumerian story is about language and control through language. It mentions Enki as the Lord of Abundance who desires to change all the languages of the people into one language, while Enlil likes to keep people confused and under control through multiple languages. The Biblical story discusses the desire of Chaos-People to go against the current rules and literally build a tower to the heavens so that they can experience freedom like the Gods. Their punishment for desiring freedom is that their one language becomes multiple languages, so they can no longer speak to each other or revolt against the Gods again.

During the days of Nimrod, the main tool that was being used for control of the people was that of language and words, with the languages being changed slightly to create Lawful systems and Legal systems of control which deceive through multiple dictionaries.

As I mentioned earlier, I believe that Tower of Babel story describes the first major revolt within Mesopotamia against Nimrod and against Enlil by the people and was actually a language revolution or Language War based on Chaos rising up against the out-of-balance situation created with too much Order.

This major revolt happened around B.C.E. 2193 and started within the city of Babel, but quickly spread across all of Mesopotamia and it is during this time in history when the Sumerians turned against their God

Enlil. The major revolution wasn't stopped until approximately B.C.E. 2110 when the lands of Sumer were finally reunited under a King called Ur-Nammu who was from the city of Ur, and who brought Enki or the <u>Chaos-God</u> to the forefront once again in Mesopotamia.

The King called Ur-Nammu was the first within Mesopotamia to give freedom to the people and formed laws to create equity in the land. He also established many laws that put into place monetary damages for harm or bodily damage to other people, instead of cruel punishments such as cutting off a hand or arm. The laws also created capital offenses for crimes such as murder, robbery and rape. It was during the time of King Ur-Nammu under the <u>Chaos-God</u> where the major Sumerian Renaissance movement started.

It was also during these times of the <u>Language War</u> that many of the great followers of Enki the <u>Chaos-God</u> came into prominence again in Sumer and overthrew most of the existing <u>Order-based</u> leadership. Most of the temples dedicated to Enlil were suddenly changed and became temples to Enki or Ea. Suddenly the minor significance of Enki became prominent in the city of Nippur and the city of Eridu. There was even a cult of Enki discovered within the city of Eridu, whose writings credit Enki with the survival of the city and its conversion into a sacred city.

It seems that Enlil and his followers had been fighting battles around the world against Enki and his followers, but this time the battle came back home to the lands of Sumer where they all started. The change however was that this battle was a revolution where Enlil and his followers were forced out of their own home-base in Sumer. You will find that once Enlil was forced out of Mesopotamia, his group would now start spending a great deal of time around the world and in other countries slowly perfecting <u>Order</u> and control through many new invisible and imaginary systems.

The <u>Language War</u> that happened under Enlil created the first brand new language and a new legal dictionary, where legal laws, rules, regulations and statues would now be created using new definitions for similar words. This is what Enlil would slowly perfect over time to help create confusion but also to use this language to enslave and control people. This perfect language for <u>Order</u> started with the Sanskrit writing system given to the Aryans from India, later perfected with the Phonetic

Alphabet given to the Phoenicians and then later completed into the modern-day English and American systems of languages.

These languages today, with their uppercase and lowercase letters are used today worldwide almost exclusively for contracts, to trade, perform banking, create treaties or even to engage in Laws of Commerce. It is important to understand that although all these systems involve the same language, Enlil has created multiple dictionaries around the same letters and words so that you may think you are hearing English spoken in a courtroom, but you are not. They use a different dictionary, so the words mean different things, and all of this has created the perfect system of control under <u>Order</u>.

<u>Abram or Abraham and his help within the Language War</u>

As mentioned earlier, Hebrewism, Islam and Christianity are the three major religions today and they all find their roots and beginning in one man and he was Abram, later renamed Abraham. Who was this man that is credited with creating exclusively those three major worldwide religions? The answer is that he was the major military leader that defected against Order and went against Enlil and the country of Sumer during the first major Language War or Chaos against Order while Nimrod was still king over Mesopotamia. Since Abram was such a significant figure in the Eternal Conflict, let's spend some time discussing him and his role in the Language War.

Abram was born in the land of Sumer, in the city of Ur, in approximately B.C.E. 2200 as we know that his birth was during the time of the reign of Nimrod and before the Tower of Babel uprising. We know this because Enlil was still being worshipped under that name during those days. As Abram grew up, he was constantly exposed to the multiple religions that were established under Enlil and their connections to the corrupt money, language and trading systems established under Order. In fact, his father Terah, who was also called Azar, was one of the high priests of Enlil within the city of Ur. His Father also owned an idolatry shop that manufactured statues of Enlil in his many forms as different Gods.

Abram worked for his Father in the manufacturing shop and at some point, during his time there, he realized how controlling and oppressive the religious and trading systems were that his father was a direct part of. So, Abram confronted his father about this, but his father ignored him and refused to listen to the advice of his son. Then one day, Abram started receiving a <u>series of visions</u> that the people should not be slaves but rather should be free and live under <u>Chaos</u>, these visions came from what he considered was angels which were probably from the followers of Enki that were trying to help him. It was after these <u>series of visions</u> that Abram then went to his father's idolatry shop and he then proceeded to smash all the God statues to pieces and drove all the customers away.

The story goes on that his father was told about this from his customers and his neighbors and so he confronted his son Abram. During this confrontation, his father Terah literally dragged Abram to the court of King Nimrod who was known by the entire family. It is there that Abram personally told Nimrod to stop worshipping the fake God called Enlil that controls and enslaves people.

During this confrontation, King Nimrod just laughed and responded by bringing out two people, one whom he sets free and the other he kills, thereby making the point that he as King can also give life or bring death.

Abram then refutes him by saying that God lets our Sun rise up from the east and so he asks the King to make the Sun rise up from the west. This made King Nimrod angry and he ordered Abram to be thrown into a fiery furnace to die. However, something happened that is unknown and so Abram was not burned up but rather survived. Whatever happened also caused his father Terah to change. The story does not specify what happened, but Abram was probably saved by the servants or angels of Enki, who also probably talked with Terah about the truth.

    I mention this guess as being logical, only because this is the moment that Terah suddenly takes Abram and Sarai and escapes from the city of Ur, leaving behind his home, his business and his religion. Since Terah was a high priest of Nimrod and was well known to all the religious and political class of people at the time, his escape was very controversial and could not be allowed to happen. They were deemed criminals and hunted down.

    I am also guessing that Terah and Abram also discovered that the normal beliefs about interbreeding within families and that bloodlines must remain pure was not a good thing, in fact they were more of a manipulation by Nimrod and Enlil, because Terah also leaves behind his only son, Nahor, who was born out of this interbred marriage with his sister Sarai. Also, Abram did continue to protect Sarai and keep her safe as a sister but from this point on, he no longer viewed her as his wife. She is called barren or without children and that is mainly because Abram and Sarai no longer tried to have children and reverted to just being siblings instead of spouses.

After their escape, the angels of Enki met up with them and told them to travel to the lands of Canaan, which as was described earlier was on the east coast of the Mediterranean Sea and is called today by the name of Palestine and Israel. They were told to travel there specifically because the land was neutral. At least regarding the battle between <u>Order</u> and <u>Chaos</u> or between Enki and Enlil.

Their journey was quickly delayed in a city that called Haaran when they were almost spotted by the followers of Enlil and they had to stop and hide. This was also where they were forced to leave their father Terah behind, as they continued their journey without him. They did finally make it to the land of Canaan where they setup an altar and a marker to Enki.

It was about this time that the revolt in Babel occurred and the Language War and revolution was in full force. This was also the time that Abram became a prominent leader of the revolution against Nimrod and against Order in general. About that time, they were then sent on a mission to go to Egypt and see the current Pharaoh of Egypt, who was on the Chaos side of the current war.

When they get to Egypt there is a strange tale of how everyone seemed to immediately notice them, so they must have been being watched for. Once they are spotted, then the Pharaoh goes out of his way and asks to see them personally. Although the Pharaoh knows that they are escaped war prisoners and fugitives, he does not turn them in, but instead gives them huge amounts of gold, silver, livestock, servants, soldiers and guards to travel with them back to Canaan. This entire story only makes sense, if the Pharaoh must have secretly supported Enki over Enlil in the battle of Order vs. Chaos and that is why the angels of Enki sent them to Egypt on this mission in the first place.

Abram and his family then returned to Canaan with all the soldiers and wealth, to form an army to fight against Enlil and his followers of Order. Abram's nephew Lot had also traveled with them from Sumer and so Abram and Lot split up the soldiers, animals and wealth. Abram stayed in Canaan to fight, while his nephew Lot took the other share and moved to the cities of Sodom and Gomorrah, where the religion of a Chaos-God called I-Am or All-Ah was already being worshipped with various high priests such as Melchizadek, as a follower of Enki.

Later in the story, as more Kings and their cities join the battle, Abram is also given more animals and more soldiers to fight against Enlil. Throughout the life of Abram, there is always many angels visiting and talking with him about battles and fights and decisions that need to be made and this is the way that Abram, as one of the main leaders, helped throughout this time of major war between Enki and Enlil.

The Language War escalates between Enlil and Enki

From this point on, Abram is appointed as the leader of the coming war and he sets up to fight against Enlil and Order that has, by now enslaved most of the Middle East and beyond. Since Abram is very familiar with the armies and tactics of Sumer, because he was raised there under a high priest, he was able to exploit all their weaknesses. The Ancient writings say that Abram shall free all the nations of the Earth and that the Earth shall be blessed because of him, which is probably why Enki put him in charge so that he may command.

From this point on, a physical battle started between the followers of Enlil and the followers of Enki. Nimrod who was also called Amraphel along with three other Kings that worshipped and followed Enlil made war against the four Kings that followed Enki including the Kings of Sodom and the King of Gomorrah. The reality is that once the Kings learned that they had been tricked into paying taxes and had been enslaved through words and contracts, the four cities that now worshipped Enki at first tried to peacefully revolt.

Those four cities had been paying major taxes and tariffs to the Kings of Mesopotamia and Enlil for at least twelve years and so under the direction of Abram, in the thirteenth year, they revolted and refuse to pay taxes and honor the trading and the contracts. That was the same year that the Kings of Mesopotamia, including Nimrod, were forced to invade the four cities including

Sodom, Gomorrah and all the cities on the south side of the Dead Sea.

Interestingly, those cities did not protect themselves or were not expecting a real war to break out so fast, that they literally lost everything they had in that first battle. The followers of Enlil took all the possessions and all the provisions of the cities. They also noticed Lot from among the people and they captured him to take him back to Sumer along with many women and other people as slaves.

Once Abram heard of his first major loss in his first battle plus the fact that his nephew Lot was now being held captive, he took three hundred and eighteen of his best Egyptian soldiers to attack and defeat this group. They then took back the possessions and the people and rescued his nephew Lot. After that instance, this is the period of Ancient history where the Language War would now escalate out of control.

If you remember back to the many earlier battles between Enlil and Enki or between <u>Order</u> and <u>Chaos</u>, both down in Peru but also over in the Indus River Valley, those battles were described in the Ancient texts are occurring with high technology, weapons and possibly rockets and even spaceships. Strangely, the ancient writings seem to describe an escalation of the Language War as the first to use nuclear weapons or something just as devastating. The Holy Bible describes the destruction of the cities of Sodom and Gomorrah as being planned out and then announced to the citizens as a form of blackmail, unless of course, the cities reversed their revolt and decided to honor the contracts and the trading and begin paying taxes again. Of course, when the cities' residents heard about their possible destruction, the people of both cities became angry and stubborn. This was normal because they had just been defeated in battle quickly and no longer wanted to listen or follow Enki anymore. (The Holy Bible, King James Version, Genesis)

The followers of Enki, described as angels, went down into the city of Sodom to the house where Lot and his wife and daughters lived and told them about the upcoming destruction. They asked Lot to intercede and to talk with the people to try and get them to leave the cities at once. This intervention did not work however, as the people saw the followers of Enki and instead they stormed the house that Lot lived in. The mob surrounding the house became so unruly that the angels used a special weapon to temporarily blind the mob and to help Lot and his family escape the city.

The very next day, Enlil used either nuclear weapons or some weapon so powerful that a rain of both sulfur and fire came down onto both cities that caused the land of the valley to rise up like the smoke of a furnace. Not just the cities were destroyed but also the valleys, the inhabitants and everything that grew on the ground. Even Lot's wife that had stopped to look back at the destruction became a pillar of white dust.

Lot and his daughters barely escaped by hiding in a cave for many days and the destruction was so intense that his daughters literally believed that the end of the world happened and that there was nobody left alive except for them. Even today, thousands of years later, both regions where the cities of Sodom and Gomorrah were located still register high levels of radiation and fallout.

The Language War helps to create Hebrewism and start Islam

The rest of the story of Abram fighting for Enki, during the <u>Language War</u>, is notable for his belief in what he calls the one true God called I-Am or All-Ah. This belief becomes the religion that Abram starts and that is specifically defined much later with his grandson Jacob, who had his name changed to Is-Ra-El based on the names of three different Gods.

There is also a time in the later years of Abram's life where he and his other son Ishmael travel to the city of Mecca in what is known today as Saudi Arabia as where Abram was preaching there against Enlil and the multiple religions and the multiple Gods of Enlil. The story goes that as he was teaching the people about the one true God, called I-Am or All-Ah, when he received a <u>series of visions</u>, again probably from the followers or angels of Enki and it was during these visions that he was given a pure white stone called the celestial stone and that was considered very special by Enki.

The angel instructed Abram to construct a special building called the Kaaba around this stone that by then had turned black and then preach to the people that they should only worship the one true God and that this building and the special celestial stone should not be worshipped but only seen as a sign proclaiming the one true God called I-Am or All-Ah. It is interesting to note that today the Kaaba that Abram builds around the celestial stone is no longer a part of the religion of Hebrewism but rather a sacred relic of the religion of Islam.

Another aspect of this religion called Hebrewism is the fight against trade by protecting certain local items, such as not using garments made of more than one fabric and not eating certain types of animals such as pigs that were not raised by local people and therefore would have been part of a Trading system that they were revolting against. A third aspect of the Hebrew religion was the fight against banking and specifically the issuance of usury or interest and debt. The Hebrews in fact had a rule that every thirty years, all debt for every individual was completely forgiven and every individual would be able to start their lives over free of all debts. This concept was called the Jubilee Year and was honored as one of the main weapons against Enlil and his followers, but most importantly against any corrupted Order-People that used banking and trading to enslave. In fact, their ancient writings collected in a book called the Torah which is also included in the Old

Testament of the Holy Bible had a statement that said to not borrow money, because, "the borrower is a slave to the lender". (The Holy Bible, King James Version, Proverbs 22:7)

The **Right to Self-Defense** is a also principle deeply rooted in Hebrewism as are many of the other freedoms, liberties and rights that come directly from **Chaos-based** thinking. In fact, the story of Abram trying to kill his only son and him following through on the killing, only to be stopped is a mythological story that directly personifies **Chaos**. The religion of Hebrewism as founded by Abram and the later subset of the religion called Judaism, Christianity but also the religion of Islam. All of these religions were founded by Abram as **Chaos-based** and that is why all three religions have been persecuted throughout the world whenever **Order** is prevalent with the world leaders.

Abram did such a great job during the **Language War** that he is renamed Ibrahim or Abraham and he is also promised many children who will be raised as Hebrews and who will continue the fight against Enlil and **Order**. One of the last aspects of the story of Abraham is that since he would no longer interbred with his sister named Sarai, he was given a new wife, one of his female servant called Hagar. Hagar then has a baby boy from Abraham who is called Ishmael. The religion of Islam teaches that Mohammed, the final prophet of their religion, is a direct descendent of Abraham through his son Ishmael. (Zeep, Ira G., C.E. 2000) ('Ali, 'Abdullah Yusuf, M.M. Pickthall, C.E. 1999).

Abraham's other son Isaac is also important but to the religion of Christianity as they teach that Jesus, their <u>Chaos-God</u> made of physical flesh, was a direct descendent of Abraham through his son Isaac. This is also important to Hebrewism, as they teach that Isaac also had a son named Jacob who was renamed Israel and he founded the Israelite nation.

Jacob or Israel, the great grandson of Abraham, had twelve sons that were originally called the twelve tribes of Israel. All of them were <u>Chaos-based</u> and supported Enki except for the youngest son Joseph, who was very <u>Order-based</u> and who wanted and loved trade and money. This son, Joseph was sold into slavery by his brothers and he ended up in Egypt where later he would become a very high member of the Egyptian court and second in command to the Pharaoh. This happened because he was very <u>Order-based</u> and became a follower of Enlil.

Later, during a time of great famine, Israel and his children eventually traveled and lived within the lands of Egypt, first through Joseph, the brother that was sold into slavery but then later as slaves in the lands of Egypt. They were also called the Hyksos, but only after their <u>Chaos-based</u> beliefs were revealed and they started to fight against Enlil. Once that happened, they were then taken over as slaves under the control of the Egyptian Pharaohs.

<u>The Hittites from Turkey were allies to Abraham in the war</u>

The Language War was just the <u>Eternal Conflict</u> acting out socially, politically and militarily and this war where Abraham fought against Nimrod lasted for some time. During those times, there was a culture that existed in modern day Syria and Turkey called the Hittites and they also worked with Abraham.

According to the Holy Bible, the Hittites supplied Abraham with wood, chariots, horses and soldiers as their contribution toward the war effort. Around B.C.E. 1590, the Hittites even formed their own army and came down into Mesopotamia to capture the city of Babylon. This all occurred, because the Hittites, just like the Hebrews and Canaanites were constantly under attack by the Enlil, the <u>Air-God</u> or <u>Order-God</u>, mostly through language and trade.

This was during the time when Enki, the <u>Chaos-God</u> was helping the people of Canaan, Turkey and Iran, where many freedoms were being enjoyed by the people and trade was limited to only what was needed. In fact, the influence of Enki can be found in the Hittite mythology where they use his real name Enki and where he was called the <u>God of Contracts</u> but only the ones favorable to humanity. Enlil, the <u>Order-God</u> was also called by name, but also by the names of Adad or Resheph and was considered to be the <u>Sky-God</u> that carried thunderbolts.

The Hittites were a <u>Chaos-based</u> organized society with a minor cuneiform <u>Logographical Language</u> up and until B.C.E. 1550 when a world-wide event occurred that changed everything at that time.

When you study the ancient history of our planet Earth - the date of B.C.E. 1550 always shows up as the time frame of major change throughout the planet - either mass destruction or major environmental damage or cultural change. Whatever happened during that time, physically altered the topography and climates of different nations and countries, literally everywhere on Earth.

If the famous scholar and writer, Immanual Velikovsky is correct, then a huge comet called Venus returned into our solar system and literally disrupted the orbits of both Earth and Mars before this huge comet settled into its own orbit as the planet we now call Venus. (Velikovsky, Immanuel, C.E. 1950)

Immanual Velikovsky discusses his ideas about Venus as a comet and a collision of worlds in a series of books, whereas he proposes that the Earth actually flipped in its rotation and our equator changed instantly. In his first book, <u>Worlds in Collision</u> he discusses this proposed global upheaval as the reason that the wooly mammoths living in the green grass lands of Siberia were instantly frozen with chewed up grass still in their teeth. He also discusses how the lands of Canaan, which were the lands flowing with milk and honey, suddenly changed and became more of a desert climate.

Velikovsky could be wrong, but whatever happened during that time-period was still a great catastrophe that changed the face of the Earth to what we see today. It was only after this great catastrophe occurred that we see a new <u>Order-based</u> religion formed around Enlil but with a new name called Teshub or Tarhunt. Teshub was the <u>Sky-God</u> or <u>Storm-God</u> that was conceived from his father Anu and is depicted as holding a triple thunderbolt in his hands. Also, within this religion, Teshub had a major conflict with a sea or water creature that was a snake or serpent or dragon, that probably represented Enki. In particular, the Hittite mythology discusses that the largest conflict occurred when the <u>Order-God</u> Teshub slayed the <u>Chaos-God</u> called Illuyanka who was in the form of a dragon and after that was allowed to take over the region.

Also, after whatever catastrophe occurred, we also know that suddenly, the Hittite power fell into obscurity for over a hundred years. Suddenly the world saw the introduction of the Aryans from India and the Phoenicians mysteriously showed up from nowhere. The Aryans came complete with their <u>Sanskrit-type</u> of <u>Alphabetical language</u> and the Phoenicians came complete with their <u>Phonetic Alphabetical</u> type of language. Both of these languages blended and eventually became the <u>Persian Alphabet</u> that over time became the modern-day English and American systems that use both uppercase and lowercase letters and today control all of the trading and banking and legal systems worldwide, under the design of International trade and banking.

Interestingly, we may not know what the catastrophe was, but we do know that there was advanced knowledge of the event, because earlier around B.C.E. 1600 to B.C.E. 1550, there was major underground cities built around the city named Derinkuyu, in the country of Turkey that helped the people in that area survive whatever happened. Some of these underground bunkers and underground cities, in fact are so huge that they could hold entire civilizations of up to 50,000 people.

It was also during this same time, that within the mountains of China were built large underground bunkers under a region that was beautiful, but suddenly became completely inhospitable to human life. These huge underground centers were built complete with very complex air and water piping that still reaches the surface today. These piping structures are called the <u>Baigong Pipes</u> and are literally hundreds of ancient iron pipes of unknown origin that go very deep into the Chinese mountains, while others go to a nearby water lake that did contain fresh water but is now full of salt water. Some of the larger pipes are over forty centimeters in diameter, but interestingly all of them are of uniform size. Today these pipes still go down into the underground bunkers and even though they are over three thousand years old, oddly they are still clean of debris with no rust.

Mo-ses as an Israelite and the Pharoah's son.

To finish up the discussion of Abraham, the <u>Language War</u> and his descendants which were called Israelites, we must move forward in time, to the time period where the Egyptian <u>Chaos-based</u> Kings were removed from power by an <u>Air-God</u> and were replaced with <u>Order-based</u> Pharaohs in their place. This would have been during this same time-period that the Hebrews or Israelites were also living in Egypt and were slaves to the Pharaoh.

The story goes that one of the Hebrew slave women became pregnant and not wanting her child to grow up a slave, she placed him in a small boat and floated it down a river. This child was found and was later called Mo-ses and strangely grew up as the son of an <u>Order-based</u> Pharaoh, who supposedly did not realize that this child was a Hebrew and not an Egyptian. Here is the most important thing and that is that since Mo-ses grew up within the Pharaoh's home, he would have had access to anything and was able to learn everything, from engineering to money to legal laws.

Later, Mo-ses discovers that he is a Hebrew and runs away and during this time alone, he receives a <u>series of visions</u> from a <u>Chaos-God</u>, who instructs him to go back to Egypt and demand from the Pharaoh that all of his Hebrew people, the Israelites be set free. This is a direct analogy to the force of <u>Chaos</u> and the political side of <u>Chaos</u> since freedoms, rights and liberties are the opposite of <u>Order</u> or control.

The story continues with a series of upheavals, plagues and widespread death, which again shows that utter <u>Chaos</u> was happening all around them. Mo-ses does finally get permission for the Israelites to escape from the slave bondage of Egypt because of all the chaotic events and possible great catastrophe that we just discussed. So, Mo-ses was able to take all the Hebrews out of Egypt and they eventually settled once again in the lands of Canaan, which were the original lands where Abraham had found neutrality much earlier, during the <u>Language War</u>.

During the trip from Egypt to Canaan however, there is another story about Mo-ses going up onto a large mountain where he supposedly talked with God and where he received what we today call the Ten Commandments or the Ten Laws. We mentioned this earlier, but the strange part of this story is that the first three of the commandments or laws are based on <u>Order</u>. (The Holy Bible, King James Version, Exodus)

The first law states that there are many different Gods, but you should never place any of those other Gods as more powerful than a certain God. The second law states that you should never draw a picture of anything at all, which would include this certain God and the third law states that you must never use the name of this certain God because you might misuse it. Interestingly, it sounds like these first three laws came from Enlil, the first of the <u>Gods of Order</u>, who was already trying to suppress his identity and name. The second law goes even further by stating that this certain God is a jealous God but will not punish the bad person, but instead will punish the grandchildren and great-grandchildren (3rd and 4th generation) of those that don't worship him, which is a very strange comment.

It is interesting to note that the name Mo-ses means drawn from water, which was what happened supposedly to him as a child but is also an indirect reference to Enki who is the God of Chaos and the Lord of Water. It is also more interesting that during the first two times that Mo-ses was able to meet directly with the Pharaoh (his stepfather), a magic act was displayed where a rod was changed into a serpent, another direct reference to Enki.

Once Mo-ses and the Hebrews could escape from Egypt, they were then constantly attacked by many other groups before they finally arrived in Canaan. However, by the time they got to Canaan, it was no longer a neutral country but instead now under the control of the God of Order, who was now worshipped by the Canaanites under the names of Molech or Baal. It was during this time that the Canaanites practiced child sacrifices and blood sacrifices which were demanded by Molech or Baal, whose symbols once again included birds, such as an eagle or an owl.

At the time that the Israelites got to the lands of Canaan, after wandering throughout the desert, that the people were begging for structure and Order. It was during this time that another interesting story arises about how some of the people were being bitten by snakes, so even though Mo-ses forbid making statues, he was then instructed to build a statue of a brass serpent and mount it on a pole, which may not have actually happened but is rather just another reference to Enki as the God of Chaos in the form of a serpent.

The fact that Mo-ses grew up as the son of a Pharaoh and received his formal education within the house of the Pharaoh, he should have been very aware of the conflict between Order and Chaos. I believe that this was the truth, because most of the religious rules that were created under Mo-ses or his brother Aaron and which are described in the Old Testament of the Holy Bible, were against the legal systems, trading, banking and especially debt. Some of these rules stated not to pervert justice, to not withhold payments from workers and to use only accurate scales when determining the value of only lawful physical money. At the time, there existed the ability to trade fabrics, seeds and animals with other nations and so to prevent this trading from happening, so more religious rules were developed such as do not crossbreed livestock, do not use mixed seeds and do wear garments of more than one fabric. One of the most important religious rules (Kosher rules) was to not trade, purchase or even eat certain animals that were not part of the Hebrew nation, so to prevent the people from purchasing or trading for them, they were deemed to be unclean animals.

Mo-ses also knew and understood that banking and debt could be used as control under <u>Order</u>, so he made many religious rules against it, by commanding that leaders never multiply horses or wives, neither greatly multiply gold or silver, which was also a direct reference to prevent today's fractional reserve banking systems, which are a part of <u>Order</u>. Lastly, as we mentioned earlier, the most important religious rule was established that every thirty years, all debt must be completely forgiven so each person could start over with zero debts. This concept was called the Jubilee Year and was the main defense that Mo-ses setup against <u>Order</u>.

King David and King Solomon

After leaving Egypt and following Mo-ses, the Thirteen Tribes of Israel (original twelve tribes minus Joseph plus Josephs' two boys) entered the lands that were occupied earlier by Abraham back when Canaan was a neutral country in the Language War. After they moved into the lands of Canaan, their nation was first governed only by judges using the few rules that had been setup under Mo-ses, but then overtime and under the influence of Order, the Israelites changed into a government under the rule of Kings. Under the Kings of Israel, Canaan would continue under stress and battle. It seems like since this specific land east of the Mediterranean Sea was the first major battle area during the earlier Language War, it will always be under attack from Chaos and Order and the Gods of both.

After rulers as Kings were established and Order grew within the nation, a small boy named David arrived who became the great King over the nation of Israel but was also very important to the religions of Judaism, Christianity and Islam. David was a small shepherd boy who was a very internal Spiritual-based or Order-based individual. He was a musician, poet and is credited with writing the entire book of Psalms within the Holy Bible. During the time that David was a boy, the nation of Israel was ruled by a very Materialistic King named Saul who is described as jealous, possessive, fearful and supposedly possessed by demons. Israel was also at war with the Philistines at this time and this also caused great distress to King Saul, so David would play music to help calm the King.

Later, David became King of Israel and always taught others that you needed prayer and internal reflection to find God from within yourself. He also ruled from the city of Hebron and then later ruled from the city of Jerusalem. King David is credited with designing an elaborate temple that would hold the Hebrew treasures, such as the <u>Ark of the Covenant</u> and this temple was created to help center the entire religion of Hebrewism around <u>Order</u>. Unfortunately, David would never see this complex court-yard and temple area constructed. His son Solomon who was appointed King after his father David was the individual who helped to construct the temple but instead of keeping it a complex structure, Solomon being more <u>Chaos-based</u>, changed it to a simple structure that would show off the materialistic urges of the King. King Solomon changed it into what would become literally one of the <u>Great Wonders of the World</u>.

The Temple of Solomon was constructed around B.C.E. 950 and instead of following the advice of his father David, who understood <u>Order</u> and the rules that were setup by Mo-ses to not get into debt. It was King David who also taught Solomon to reject the trading systems and money systems of other countries, but Solomon, as King, decided to deviate from his father's plan and construct it out of the best wood and finest gold. To do this, Solomon had also agreed to trade with other countries and so as the Phoenicians made agreements with him, it soon become clear that even though they arrived from nowhere, they were the ultimate traders at the time, based around <u>Order</u>. King Solomon made agreements with the Phoenicians and those included accepting banking, debt systems and Legal agreements of international trade.

The Temple of Solomon was constructed out of the best cedar wood imported from Lebanon, which then was overlaid with pure gold. The Phoenicians also provided all the skilled craftsman that were needed, of course at a cost and under the direction of their best craftsman called Hiram Abif. The external doors to enter the temple and the internal doors were huge and also overlaid with pure gold with a huge veil made of many different colors of the finest linen was hung from one wall to the next. There were two rooms inside the temple called the Holy Place and the Holiest of Holies, which was the room to house the <u>Ark of the Covenant</u>. Both rooms were lined with imported cedar and had multiple statues of Angels or Cherubim, palm trees and flowers, all overlaid with more pure gold.

Outside the temple were two huge pillars, named Boaz and Jachin and they were about 27 feet tall and cast from pure brass. They were no longer used to record or determine the times and the seasons during one year like they did in the past. Instead they were made as hollow phallic symbols or as twin towers, like the Twin Towers that were already located in the cities of Tyre, Byblus, Paphos and Telloh - and thousands of years later on Manhattan Island in New York, USA.

King Solomon was considered extremely wealthy, but only because he bought in the money systems that made him wealthy, in exchange for adopting the <u>Order-based</u> banking and money systems plus instituting a new tax system against the Hebrew people. The Phoenicians were also responsible for creating the new Legal systems and the new huge Judgment Hall that King Solomon used as he became the new judge of his people. Interestingly, Although King Solomon was considered extremely wise, that was only because of his adoption and construction of the Legal Codes and court systems within the nation of Israel that preached fairness but restricted freedoms, liberties and rights of the people.

King Solomon quickly developed the bad habit called the Power to Control and the Power of Money as he also became addicted to women as he had over 700 wives and 300 concubines. Some of his wives were foreigners, including the daughter of a Pharaoh and women from the cities of Moab, Ammon and many from the Hittites. As King Solomon was tricked by the force of Order and manipulated by the Phoenicians, he ultimately was the King that introduced both idolatry and slavery into the nation of Israel. He may be talked about by historians as one of the greatest and wisest Kings but in reality, he was actually personally responsible for helping to enslave his people.

Israel tries to move back to Chaos and is attacked

King Solomon knowingly or unknowingly enslaved the nation of Israel and then after his death, his son called Rehoboam took over as King and continued down the same path. King Rehoboam took the trading, money and legal agreements that his father had setup and increased them tenfold. He imposed massive taxes on the Israelite people and then went further and took away even their right to be a self-sufficient people under the direction of the force of Order.

This led to sudden rise of Chaos in B.C.E. 930, when the nation of Israel had become so oppressed, that a Civil War broke out among them. By that time, the country of Israel had split up into two nations, with ten northern tribes of Israel rejecting King Rehoboam and moving toward Chaos, while the three southern tribes of Israel continued to accept Order, mainly because they which were the tribes that ran the trading, banking and legal systems. The southern tribes were the tribe of Benjamin, the tribe of Levi, who were the High Priests at that time and the tribe of Judah, from whose name these southern three tribes would come to be known as the Jewish nation. This civil war between the northern and southern tribes of Israel lasted throughout the rule of King Rehoboam.

This was the same time period where the supposed <u>God of Order</u> and his followers also returned back to Mesopotamia and to both Sumer and Assyria and in particular the city of Babylon. In the year B.C.E. 720, the Assyrian King called Saragon II began helping the three southern tribes of Israel to end the civil war against the ten northern tribes of Israel, by attacking them and driving them out of the lands east of the Mediterranean Sea. This time, ten northern tribes had either escaped into Europe or had been deported back to Assyria or Sumer. These ten northern tribes today are referred to as the Lost Ten Tribes of Israel.

The high point of <u>Order</u> from a historical basis occurred in the Middle East around B.C.E. 605, when a new King called Nabopolassar created what would be called the Babylonian empire, by taking control of both Syria and Phoenicia and after the death of King Nabopolassar, then his son King Nebuchadnezzar also continued his conquests, by taking control of Palestine and the city of Jerusalem in the year B.C.E. 597. The capture and control of Palestine is an important event for the religions of

Hebrewism, Christianity and Islam, because this time when the city of Jerusalem was finally captured once again, both the city and the holy Temple of Solomon was completely destroyed, plus almost all the Jewish people from the three remaining southern tribes, were taken and deported back to Babylon as slaves along with all of the Jewish leaders.

    King Nebuchadnezzar then took control of the countries of Tyre and Egypt before finally returning to Babylon and rebuilding the city with aqueducts, temples and water reservoirs. He also started building very complex ziggurats and complex buildings and these construction projects that were completed under Nebuchadnezzar was massive. He also built underground passages within the city and a huge stone bridge over the Euphrates River that was supported on asphalt covered piers that were engineering to reduce the upstream resistance to the flow of the river. King Nebuchadnezzar also restored the lake of Sippar and opened a new port on the Persian Gulf to allow more trading. His greatest projects were that of the <u>Hanging Gardens of Babylon</u> and the <u>Ishtar Gate</u>, which is one of the largest gates leading into the city of Babylon.

Aryan migration into the Indus River Valley and out of India

Around B.C.E. 1700 there came a new God, within India, who was called Varuna and was depicted as a <u>Sky-God</u> or the <u>Law-G</u>od, who job was to fight, control and try to regulate the cosmic <u>Chaos</u>. Varuna was probably the same God called Asura within India's <u>Rigveda</u> writings, as he was also the Sky-God that became the guardian of all law and was considered to be the cosmic <u>Order</u>. Also, in many writings, Varuna was equated with the mighty warrior called Indra who was also described as the Sky-God of Thunder, who traveled and lived in the atmosphere of Earth.

Strangely, around B.C.E. 1550 within southern China and northern India, there were many stories of a migration of white creatures into the country of India. These white creatures were called Aryans and although there is very little physical evidence for this happening, the stories coincide with the decline of the Harappan culture within India, the rise of <u>Order</u> in that region, the sudden creation of the <u>Alphabetical Language</u> called Sanskrit and the expansion of white people out of India and into Persia (Iran), Turkey and the European countries. This expansion allowed the new Sanskrit language to be carried westward through parts of the Middle East and into Europe, where today it is acknowledged that there are many similarities between this new Sanskrit language and the later European languages, especially after it combined with the Phonetic Alphabet of the Phoenicians.

Interestingly, Sanskrit writing is an Alphabetical Language that came out of nowhere and consisted of between 36 and 48 sounds and its writing has letters and vowels composed as words and sentences. This Sanskrit language was not taught for the common people, but instead was considered a refined or perfected manner of speaking or writing. In fact, if you had knowledge of Sanskrit at that time, then you were considered to be part of high class or part of the upper social order. This Sanskrit Language was mainly taught to the traders, bankers, government leaders, religious leaders and the higher castes of people.

One form of Sanskirt writing was called the Vedi-Sanskirt or simply the Veda and was mainly four different bodies of writings from around B.C.E. 1500 consisting of hymns, dogma, prayers, stories or spells and incantations that were created by very Spiritual Leaders that needed to control their Materialistic followers. The religions of Hinduism fought against these writings for a while, but eventually did accept them but only as writings to study and contemplate internally. The religions of Buddhism and Jainism and most of the non-Brahmin Hindus never accepted them as sacred texts even to this day.

One note regarding this section, that when discussing Sanskrit writing or more importantly the supposedly Aryan people, it is important to understand that some of these ancient stories and ideas from the writings was to later be perverted into very racist ideas, that are not based in fact at all. So, keep in mind that using the word Aryan to promote racism is just stupid and not even logical at all.

The Phoenicians as Traders under Order

Around this time, there were Sea-People called Phoenicians that are said to have rose up from within the Canaanites but rose up seemingly from nowhere. They created the first major Maritime Trading systems with boats all along the Mediterranean Sea starting around B.C.E. 1550 and thrived all the way to around B.C.E. 300

and created many monopolies on both Trade and Money. This was another culture to rise out of nowhere and automatically have an extensive Alphabetical Language which was also used for their Legal Laws that governed their International Trading schemes. The Phoenicians were the first traders to force all countries that they traded with, to adopt their new language and this caused their language to spread throughout all the Mediterranean countries and even into northern Africa. Today this early Phoenician Language is considered the ancestor of all modern Alphabetical Languages used today including the English language that is used worldwide for International Trade and Legal Law systems.

They were also one of the first Order cultures to organize their government and leadership using cities under the rule of states and then under the rule of the main government like we do today. They were also the first culture to create a world economy based on International Trade and were among the greatest traders of their day. Part of the reason was their discovery of a special dye that makes the royal purple color, which was created by the Phoenicians from the Murex sea snail. This dye made them so much money that they could afford a huge Navy as well as their huge trading systems. They also worshipped Molech and Baal as <u>Air-Gods</u> and practiced child sacrifices and animal sacrifices as required through their <u>Order-based</u> religions.

<u>The rise of Zoroastrianism</u>

In the country called Persia or what today is known as modern day Iran, a man named Zoroaster was born in the Spitama clan where he grew up and worked a polytheistic religion as one of their High Priests. The society that he lived in was very <u>Order-based</u>, where even the government was also the religion of the people and where they had a very oppressive class and social structure that enslaved and controlled all the people.

According to their holy book called the Avesta, when Zoroaster was approximately thirty years of age, he went down to the Daiti River and it was there that he received the first of a <u>series of visions</u> from one of the six immortals, called the Vohu Manah. After receiving these visions, Zoroaster rejected what was then the religion that he was preaching in and fought against the trading and banking systems of the Princes and other High Priests of his day. The <u>series of visions</u> that he received transformed his view of the world and he then tried to teach others what he had been shown. At the time, the Persian religion taught that there was seven divine Gods, and the highest was called Ahura Mazda, the Order-God, whose symbol was the sky and specifically the winged disc. Under his direction, his son called Atar, fought against what was described as the lesser Chaos-God called Azhi Dahaka, who was known as the great dragon. To Zoroaster, the Order-God, called Ahura Mazda was considered to be the God of Reason. To worship this God was considered wonderful, but only as an internal religion that acts only through the mind.

Zoroastrianism today is still found in Iran but mainly concentrated in the country of India because around the year C.E. 651, most of the followers of Zoroaster had to escape when the Muslim armies invaded and took over most of Persia.

It is interesting to note that the followers of Zoroastrianism still believe that the highest God called Ahura Mazda was the God of Creation, which had two different conflicting sides called Asha or Order, while the other was called Druj or Chaos. It is also believed that when nature itself was raised up and created, it was done so under the symbol of the serpent and that is why in Persia, you will find many temples that have been built to the serpent tribe, which created the Earth and all of humanity.

The religion called Zoroastrianism states that one of its understanding is that achieving balance between <u>Matter</u> and <u>Spirit</u> is critical and this balance happens with active participation in life of both positive mental thoughts and positive physical deeds, as both of them is the only key to true contentment. Zoroastrianism does believe in a final renovation of the world, whereas all people will be judged by God but instead of punishment, there will be two judgments and the first will be that of Menog or <u>Spirit side of humanity</u> and the second judgment will be that of Getig or the <u>Material side of humanity</u>.

<u>Brotherhood of the Snake based on Chaos</u>

Before moving away from the history of the Middle East, Egypt, Sumer or Babylon, it is important to note that out of all the animals that you can find on Earth, there was none as noticeable or as sacred as the snake or serpent throughout Middle Eastern history and even our ancient worldwide history. There was even a secret society established called the <u>Brotherhood of the Snake</u>

which was buildup to help people learn what they called the sacred knowledge, truth and freedom, which was all information based around <u>Chaos</u>.

The ancient Egyptians had writings that stated that the <u>Brotherhood of the Snake</u> was a great organization that opposed the enslavement of people. This organization had to go underground however, when the <u>Chaos-based</u> government of Egypt at the time had leaders that were Kings, suddenly had new leaders called Pharaohs that supposedly came down from the Gods to control humanity. Toward the end of the <u>Order-based</u> rule of Pharaohs in Egypt, this specific secret society expanded by sending out members throughout all the civilized world, establishing branches and memberships. What the <u>Brotherhood of the Snake</u> did this at that time is unknown, but it may have something to do with preparing for eventual end of the <u>Order-based</u> Pharaohs at that time.

Chapter 12 - HISTORY OF ANCIENT CHINA TO ATLANTIS

Order and Chaos within individuals, society and history

Before we move further into the ancient history of Earth, you may be asking yourself whether the two forces of <u>Chaos</u> and <u>Order</u> that were described in the previous chapter are what is responsible for molding societies and cultures or maybe it is just the repetitive nature of humanity that forms a simple coincidence? After all, isn't there always people that want to control others and take power for themselves and isn't there always people that want freedom and independence, and those two types of people are just a normal part of humanity?

This is a very important question, because it is true that the repetitive nature of leaders and politicians is to rise up and take power and control, suppress their citizens and then later the citizens will rise up and take back their freedom, whether through a revolution or broad resistance and all of this is a natural part of humanity and is a cycle that repeats itself, over and over again, throughout history. But I would also argue that if that the pattern is not all there is to it. In reality, the two forces of <u>Chaos</u> and

Order were at work, because there was much more occurring then just coincidence.

As we mentioned earlier, if you review the Scale Chart on page two of this book, you will see entire list that shows all the different aspects that happen when Order or when Chaos moves to the forefront. When there is an out-of-balance situation between the two forces, and one of them moves to balance out the other, they become as the stimulating factor that works behind the scenes and that forms or changes societies, cultures, and history.

When you see the force of Chaos change or mold a society, you see more than just people wanting freedom, you see people suddenly use the snake and water as their symbol or banner, laws being reduced down to only a few simple laws, lawful money established based on actual commodities; the formation of simple limited governments; pyramidal structures being built, and all the other aspects of Chaos as it rises up.

When you see the force of Order change or mold a society, you see more than just people taking control and dominating other people, you see people use air and birds, more often the eagle, as their symbol or banner; massive amounts of laws being created; money changed to Legal tender, fiat or debt-based currencies; the formation of very complex governments with lots of levels and agencies; the construction of very complex buildings; and all the other aspects of <u>Order</u> as it rise up.

Reviewing this <u>Scale Chart</u> shows that <u>Chaos</u> and <u>Order</u> are not just coincidence or a repetitive part of humanity, but rather they are incredibly active forces that not only form all the subatomic structures inside everything and each person individually, but also do also exert their forces upon every aspect of the universe to create different societies and different cultures collectively.

<u>China was established under Chaos but struggled throughout</u>

Moving back to the study of history, you will find that China was one of the greatest nations to ever be setup based on <u>Chaos</u> around B.C.E. 4000 to B.C.E. 3000 and since they are very isolated from the rest of the world, with great deserts to the north and west, vast mountain ranges to the south and the Pacific Ocean to the east, China was one of the few countries that has the ability to truly understand the internal struggle between its Chaotic accepted roots and the force of Order.

Although as a country that was first established under Chaos, China still has had long time periods where the force of Order would take over and then decline to revert back to Chaos. In fact, whenever the force of Chaos would exert itself upon the nation of China, it would do so quickly and with the full acceptance and understanding of its people. The struggle between the two forces has lasted so long and the time has been so unrestricted that its people understand the need for balance and duality. This understanding of balance and acceptance of Chaos is one of the main reasons that the snake or the dragon, within all of China, is still revered today as an important motif and is the major character in literally hundreds of folk tales, legends and myths. You may not be aware that China is also the country that has more ancient pyramids and pyramidal structures than any other country on the planet.

Even today, the Chinese people still use their ancient Logographical language where one symbol represents one word or concept and this language also reveres and even worships snakes, serpents and dragons. Their <u>Logographical language</u> is one of the most extensive in the world with almost 9,000 different symbols representing almost everything and it seemingly started out of nowhere.

The ancient history of China from around B.C.E. 3000 was said to have been established by the <u>Three Sovereigns</u>. These were a group of demigods known as God-Kings that established China under the idea of imparting knowledge and language to all the people. It is interesting to note that usually when <u>Chaos</u> is involved, there seems to almost always be a trinity or triune of relationships being present. This ancient story about the beginning of China also involved what is referred to as the later <u>Five Emperors</u> that worked with the three God-Kings to teach the people farming, home construction, the manufacture of cloth and silk, the calendar, medicine, language, as well as the introduction of fire for heating and cooking. It was only after the reign of the three God-Kings and after the death of the last of the five Emperors named Shun did their first dynasty, called the Xia Dynasty emerged under Yu the Great around B.C.E. 2070.

There have been many archaeological discoveries directly attributed to the Xia Dynasty, such as around the city of Yanshi and in Henan, but many scholars still doubt the stories from this time-period. With that in mind, one of the greatest success stories that came out of the Xia Dynasty was the construction of large canals under the direction of Yu the Great, which helped to stop the flooding of the Yellow River but also helped to increase agricultural production for all the people. This construction was very difficult and had many setbacks during its thirteen-year construction period.

The Xia Dynasty is important as it marks the establishment of <u>Chaos</u> for almost 1,400 years before the Shang Dynasty was to arise and take its place around B.C.E. 1600. The Shang Dynasty was the first time that the force of <u>Order</u> intervened in the lives of Chinese people. We know this because while the Xia Dynasty is known as the representation of water, the Shang Dynasty is known as the representation of fire, birds and the air.

The Shang Dynasty was the first to establish <u>Spiritual</u> religions that revolved around belief in a High God known as Di; in the afterlife, as is demonstrated by their royal tombs; powers of nature, specifically the wind and the sun; the power of deceased ancestors, that had power over them from the grave; and divination and rituals. Animal and human sacrifices were also a part of their main beliefs, which usually always happens when Order takes control.

Around B.C.E. 1046 the Zhou Dynasty took over and Order continued and eventually expanded into the Qin Dynasty. These were the most controlling and dictatorship like governments that every existed and they literally tried to remove all the Chaos symbols and elements of Chaos from the Chinese culture. The Zhou Dynasty or Empire was the first in China to start taking away the rights of the people and to introduce their mandate from the heavens or as mentioned earlier, was called the Divine Right to Rule that supposedly given to the new Order leaders by the Gods.

The Qin Dynasty started around B.C.E. 221 and it was the first Order government to create an extensive legal system of laws that was controlled by a centralized legal form of government. The Qin Empire demanded that everyone give strict allegiance to the new legal codes and the government ruled with absolute control and power.

During this time, the money and trading systems were updated, tightly regulated and controlled, with new taxes and even using a new currency of China. So strict were the legal codes during the Qin Dynasty that even the measurement and sizes of axles on wagons were regulated. Of course, all books and scrolls not approved by the government were censored and any political speech or any political uprising was immediately punished by death. <u>Order</u> and <u>Order-based</u> governments need to censor speech and writings to remain in control. All scrolls or writings that were not approved by the central government were burned or destroyed.

In B.C.E. 722, <u>Order</u> became overwhelming in China and <u>Chaos</u> started to take back over, when the Zhou Empire no longer had control of the people and the government was quickly decentralized by the Chinese people into hundreds of smaller states and cities. In other words, <u>Order</u> became so strong and out-of-balance that it literally created both internal and external pressures and caused the kingdom to implode. It was during the decline of <u>Order</u> and rise of <u>Chaos</u>, during the last years of Zhou Empire, that China began many pyramid construction projects. It was also during that same timeframe that <u>Chaos</u> gave freedom to millions of people, suddenly hundreds of schools of thought and philosophy arose within China, such as Confucianism, Taoism and Mohism. This rise of <u>Chaos</u> from B.C.E. 722 until B.C.E. 476 is known as the Spring and Autumn period in China.

Interesting, this rise of <u>Chaos</u> within China was not isolated but happened at a similar time-period in many places worldwide, including Greece where their famous Republic rose up and included the many schools of Greek philosophy and thought that included Plato and Aristotle.

Around B.C.E. 202, the Ham Dynasty arose under <u>Order</u>, and pushed through more trading and money systems, opening free trade routes through Mongolia and what was then the Roman Empire in Italy, but again <u>Order</u> was pushed aside by the Chinese people. The decline of the Ham Dynasty happened quickly as by A.D. 220, <u>Chaos</u> again took its rightful place in China and the philosophy of Confucianism became a religion and a huge following. It was also during this decline of <u>Order</u> that thirty-two additional pyramids were built within China.

If you follow the rest of Chinese history, up to today, then you will discover the same back and forth struggle between Chaos and Order throughout all the Imperial periods and different Dynasties all the way to today. The establishment of the Chinese Republic form of government in the year C.E. 1912 under Chaos disappeared under the establishment of the People's Republic of China, which holds power even today. This continuous back-and-forth struggle throughout its long and vast history gives the China an upper hand, regarding the true understanding of and Order and Chaos that still continues today.

China is the one country that completely understands the need for balance between the two forces. Regarding balance, there is a Chinese God named Shangdi (similar to the Sumerian God called Anu) who supposedly ruled over two lesser Deities including an Air-God and a Dragon-God (similar to the Sumerian Enlil and Enki). Shangdi main task was to control and balance both of the two lesser Gods and this idea is still very prevalent in China. China was also one of the first nations to setup a balance between Order and Chaos through the establishment of Warrior Priests such as monks, ninjas and samurai soldiers. These religious leaders were the first to combine and balance their internal spirituality with their external physical skills of being a warrior. The

Buddha and the religion of Buddhism is also still revered strongly in China because of the belief that the Buddha was the one that was able to balance Order and Chaos and is shown in an Chinese illustration called the Maha-samaya Sutta where the Buddha is shown as the only individual that created a peace accord between the Naga serpent gods and the Garuda eagle gods.

The Main Religions of China - Confucianism, Taoism and Mohism

To understand some major worldwide religions that were founded within China, you only have to look within the many time periods that both Order and Chaos struggled back and forth within their government and within their society. As you follow history, you will discover that each time a repressed society, that is suffering under too much control and Order, is suddenly opened up under Chaos and given freedoms, rights and liberties, they will almost always make huge gains in philosophy, thought and religion as well. Within ancient China, each time Order became suppressed and Chaos expanded, suddenly huge intellectual movements arose, namely Confucianism, Taoism and Mohism.

In China, around B.C.E. 722, as the earlier <u>Order-Based</u> Zhou Dynasty was quickly being decentralized by the Chinese people, with much help from the <u>God of Chaos</u> and his followers, their <u>Logographical Language</u> was reinforced, and the Chinese people suddenly built their first and second pyramids.

It was also during the time-period of <u>Chaos</u>, that the religion of Confucianism was developed by a wise man named Confucius (B.C.E. 551 to B.C.E. 479) who was a teacher, philosopher and a politician. The philosophy and political views of Confucius emphasized morality of the government and each individual. In fact, one of his most famous principles was that of <u>Whatever goes around will eventually come back around to you</u>, actually described by him as, <u>Do not do to others, what you do not want done to yourself</u>.

Interestingly, Confucius spent much of his early years as a Minister of Crime, within the government, trying to teach the value of proper conduct and righteousness to all the people and different cities. His original goal was that of setting up a legitimate government in the area that would foster freedoms for all people. At first, he had an idealized vision of creating a <u>Chaos-based</u> centralized government that would reflect his teachings, but he eventually gave up on the idea as the other leaders he worked with, ended up resorting to a violent revolution of which he disapproved of.

Confucius eventually resigned from politics completely and focused instead on his teachings and writings. His system was based on many of the same principles that Enki taught earlier, in ancient Sumer, which was internal study and development of a sound judgment from within instead of from other people. He also demonstrated that God is within each of us and not external and that all humans have intelligence which makes them more important than either property or animals. His <u>School of Thought</u> taught that gaining knowledge and achieving internal balance is critical to learning and that righteousness should always be performed even if it is against your own self-interest.

Taoism also arose during this same period of Chinese history and is both a philosophy and a broad religion. The term Tao means the source and essence of everything and is a mixture of two things, which I refer to as the <u>Eternal Conflict</u>. The philosophy of Taoism also reflects the earlier teachings of Enki and is individualistic and was never an institution. It slowly evolved over time from several different schools, but basically focuses on simplicity, balance and harmony between <u>Chaos</u> called matter and the <u>Order</u> called spirit, focusing also on action through inaction.

Within Taoism is the principles of longevity or immortality but the symbol and teachings of the <u>Yin and the Yang</u> is most important. The <u>Yin and the Yang</u> literally means dark and light mixed and teaches the fact that all things are dual natured, but not necessarily opposed to each other but rather they are complementary opposites. This principle teaches that there are two sides to everything but although they are opposites, they also interact within everything and therefore cannot exist alone. For example, cold cannot exist without heat which complements it. Which is the same as visual light that cannot exists without darkness to compare it to. Taoism rejects the idea of good against evil completely and views it only as moral judgments made by people and society. The most important is the balance that is needed between <u>Chaos</u> and <u>Order</u> which is universal.

Both Confucianism and Taoism slowly brought to light a new concept called Mohism which started out as the concept of <u>Universal Love</u>. The governmental system within Mohism is sometimes called Meritocratic but is a system that focuses on giving freedoms, liberties and rights to the people to allow talented people to thrive by the government getting out of the way with limited regulations and restrictions. Mohism also teaches that all governments and leaders must have values and standards that originate from the Heavens and they must hold the <u>Laws of God</u> above all others. This system is a minor governmental system that should be a part of every Republic form of government and why governments should never regulate any religion, but instead be constantly reminded that freedoms, liberties and rights come from God.

Mohism also teaches that a perfect government must therefore love all the people benevolently by securing their freedoms, liberties and rights, because if they don't then they will slowly find <u>Seven Disasters</u> happening within their country. The first of these is described as the government spending money on all the wrong things and the country's defense will be neglected. The second is that once greed and unrighteous gripped the government, from within, then the neighboring countries will not be willing to help anymore. The third of the

<u>Seven Disasters</u> will be that the people living within the country will become lazy or engaged in unconstructive work will be rewarded for doing nothing. The fourth is that there will be massive creation of laws, to the point that the laws, rules and regulations will become too heavy and will develop fear in the people, who will then stop caring about the country or government and look out for only their own good. The fifth is that the leaders will then have a mistaken view of their own ability and the country's strength. The sixth is that the people will no longer trust the government and whoever the government trusts will no longer be loyal. The loyal people that still care about the country will no longer be trusted though and will even be arrested for being loyal. The seventh and last of the <u>Seven Disasters</u> is that punishment for crimes will become so lax that criminals will no longer be afraid of committing crimes and the leaders will no longer be able to carry out their work. Food will then become scarce.

<u>Mayan civilization created under Chaos in Mexico</u>

Like the sociological aspects of the Chinese culture under <u>Chaos</u>, in Central America, the Mayan civilization was also established under <u>Chaos</u> around

B.C.E. 2,000 and they reached their highest state of development around C.E. 250 to C.E. 900. The Mayan civilization was known for its fully developed written <u>Logographical Language</u>, extensive art and architecture. They had very extensive astronomical and astrological systems with very extensive calendars. In fact, the influence of the Mayans can be found through Central Mexico to Honduras, El Salvador and Guatemala along with the influence from another similar group called the Toltecs.

Both the Mayan and Toltec civilizations were very proficient at building pyramids and they literally created hundreds and hundreds of pyramids all over Central America including the great pyramids and city of Teotihuacan, along with great pyramids and city of Tenochtitlan.

The Mayans constructed the world's largest pyramid by volume, which is the world's largest monument by volume and it was called the <u>Great Pyramid of Cholula</u>, located in Cholula, Puebla Mexico which stands at over 150 feet tall with a base that is 1,300 feet by 1,300 feet. Its total volume of stone construction is approximately 4.45 million cubic meters. According to its early mythological story this pyramid was built after the worldwide flood by a giant named Xelhua. The giant Xelhua built and dedicated the pyramid to the <u>God of Chaos</u> called Kukulcan or Quetzalcoatl and who was considered the great feathered serpent God.

Interestingly, the Mayan culture was an ancient play called the Rabinal Achi, which could also be compared to the ancient Sumer story of the <u>God of Chaos</u> called Enki and the <u>God of Order</u> called Enlil. The Rabinal Achi play has two main brothers that are also princes and that are in conflict with each other. One brother is very positive toward people in general and he is called the Rabinal Achi or the Prince of Rabinal. The other brother is very negative toward people and he is called the Kiche Achi or the Prince of Kiche. In the play, the negative Prince of Kiche is captured and put on trial for trying to kidnap all the children of the positive Prince of Rabinal. There are many other characters within the play including two groups with the sacred number that represent the warriors and they are thirteen jaguars and the thirteen eagles. This play, within the ancient Mayan culture, was so significant that even the modern world declared it a <u>Masterpiece of Oral</u>

and Intangible Heritage of Humanity in the year C.E. 2005.

The significance of the God of Chaos within the Mayan culture cannot be overstated and as mentioned, he was called Kukulcan to the Mayans and considered a God that was positive toward humanity. This same God of Chaos was called Quetzalcoatl to the later Aztecs, but they viewed him as a negative God, because they were founded under Order. Kukulcan or Quetzalcoatl was described as a white boy that was initially born as a snake or feathered serpent and who helped bring the written word (as a very extensive Logographical Language), communication and peaceful trade without money or debt to the many different societies within Central America. The boy later grew up and became a winged serpent once again, but this time he had a beard and he could fly as far as the Sun. The primary religious center of Kukulcan was the area of Cholula where the world's largest pyramid by volume still stands today.

In some cultures, the iconography of the feathered serpent cult throughout Central America, also related Quetzalcoatl to both the creation of humans and the creation of the planet Venus. Quetzalcoatl or Kukulcan was considered the <u>Chaos God of the Morning</u> or the <u>Morning Star</u>, while his brother Xolotl or Kukmal was considered the <u>Order God of the Evening</u> or the <u>Evening Star</u>. Interestingly, the <u>Evening Star</u> was also considered the planet Venus, which was also called <u>Lucifer</u> to the people of Europe, during this same time-period.

Aztecs created under Order in Central America

Around the year C.E. 1300, another group would be created within Central America, but this time under <u>Order</u>. They were originally a group of wandering nomads called the Mexicas but would later be called Aztecs. Some stories mention that the <u>God of Order</u> wanted to create an enemy of the Mayans and Toltecs, so he trained them in engineering, construction, science and astronomy. The long-term goal of the God of Order was to get this new rival, the Aztecs, to eventually attack and destroy the Mayan and Toltec cultures.

According to the ancient stories, in approximately the year C.E. 1323 the wandering nomads or Mexicas were suddenly exposed to a <u>series of visions</u>, that showed an eagle sitting on a pear cactus and the eagle was eating a snake. To these Mexicas, this was a very fearful vision, because the stories of the God of Chaos called Kukulcan or Quetzalcoatl (who was the snake) was well known to all the people, including the Mexicas. Then the vision showed the eagle completely destroying the snake, while great technologies and strange weapons surrounded it. The Aztec legends describe that the series of visions ended with an explanation that they should stop wandering. The vision ended by showing them a strange location where they were to build their new city, which strangely was a location where no city should have ever been built, because it was on a swamp island in the middle of a lake.

In the year C.E. 1395 the Mexicas modeled their island capital in the lake called Texcoco after the huge city of Teotihuacan. They built extensive aqueducts, roads and bridges for their city and this group of wandering nomads was able to build their city to float on the surface of the swamp water. The technology that they used rivaled that of ancient Rome constructions during their peak.

For the next forty or fifty years after they settled down in their city, the Mexicas became a huge military power that repeatedly attacked other cities and assassinated other leaders. By the year C.E. 1428 other leaders joined up with the Mexicas and their coalition became what we today know of as the <u>Aztec Triple Alliance</u> or simply as the Aztec Empire. Since the Aztec Empire was constructed under <u>Order</u>, they started with an Oligarchy type of government organized in city and states ruled by Kings. They soon began their system of expansion through conquest and war, in which all their conquered lands had to pay tribute of wealth and slaves to the Aztecs. Of course, the same as all other <u>Order</u> cultures before them, both blood sacrifices and child sacrifices were demanded and ritually occurred under the direction of the High Priests. The Aztecs had a huge commercial center that used Cacao beans and standardized lengths of cotton cloth as the money system. Trade was demanded as the self-sufficiency of the people was not allowed. In fact, selling your labor as a commodity was not allowed under Aztec rule, but you could sell yourself or your children as slaves or as tribute as a child sacrifice.

The Aztecs were also some of the first professional merchants and traders to travel between other Central American cultures and they would sometimes use small gold statues as the commodity or as money within these various marketplaces. To protect their trade systems, various legal systems were put in place was used by the traders and the merchants and could also be used by judges and scribes or lawyers to settle any dispute.

According to Aztec legends, the Aztec empire was run by very extremely mean people that followed the orders of the <u>God of Order</u> directly. The <u>God of Order</u> was called Xolotl or Kukmal and he was known as the brother of Kukulcan or Quetzalcoatl. Of course, Kukmal is shown within all the Aztec, Toltec and Mayan cultures as either an eagle or a jaguar and Quetalcoatl as a white, bearded or feathered serpent or snake.

The Aztecs were very successful at conquering many lands, starting multiple wars, and destroying many cultures. After the Aztecs became a dominant force in Central America, suddenly the concept of serpents and snakes as Gods were considered negative and incorrect.

Ancient Greece and their language

Moving away from Central American and moving over to Europe, we find that the earliest forms of ancient Greek writing and language, established under <u>Chaos</u>, were picture symbols as a type of Logographical Language. But by B.C.E. 1900 the language was already starting to be replaced, where the pictures became changed to represent syllables. By B.C.E. 1550 when the traders known as Phoenicians arrived, they started to take control of the countries around the Mediterranean Sea with their new language and trading systems.

However, the country of Greece did not change fast enough for the Phoenicians, plus they would not accept the new trading systems from the <u>God of Order</u> whom the Grecian people called Zeus, So, the country of Greece was exiled and purposefully plunged into a dark age, that lasted almost 300 years from between the years B.C.E. 1100 to B.C.E. 800. It was only after the country of Greece finally accepted the trading systems and banking systems from the neighboring country of Phoenica was the country finally allowed to advance again.

To accept trading however, Greece was forced to accept the language of the Phoenicians which consisted of both letters and an alphabet. Of course, the first use of this Alphabetical Language was to record all their new Laws and to write the stories of the new Phoenician religion of the <u>God of Order</u> called Molech. This new religion was also changed slightly by Greece, into a religion that consisted of worshipping of the ancient

Sumerian God called Anu, father of Enlil and Enki, but was now to be called Kronos or Chronos. This new religion of Kronos required a ritualized sacrifice of children by fire to this new God, whereas a huge statue of Kronos was made from bronze with arms outstretched and then children were burned alive by laying them on the outstretched arms of this statue with a huge fire beneath.

Overtime, the <u>Brotherhood of the Snake</u> organization, as mentioned earlier arrived in Greece and began to help the people understand what was happening. The Grecian people then eventually realized that they were being put in slavery with the trading, money and debt systems, so they simply changed their language completely by adding additional letters called vowels and added them into each of their current words. By adding vowels to the current words, suddenly the old contracts and Laws were not written correctly and were instantly made invalid. This eliminated the ability and force of the legal laws that were already written, as they were no longer the correct language.

The country of Greece then made one more even more drastic change to their language, which would seem extremely strange, unless you understand the <u>Eternal</u>

<u>Conflict</u> and the language battles that continually occur through history between the forces of <u>Chaos</u> and <u>Order</u>. You see, the written language that was being written during that time was written and read from the right side to the left side of all documents. However, suddenly in the B.C.E. sixth century, the entire country of Greece switched to reading and writing their language from the left side to the right side and strangely the entire country happily agreed and implemented this change.

Once again, this change in language freed them from all their debts and contracts, because the current documents no longer fit the current new writing standard. This drastic change to their language happened through <u>Chaos</u> and resulted in the removal of <u>Order</u> from the scene and resulted in incredible prosperity for Greece and the revival of science, medicine and philosophy throughout that country.

For example, Pythagoras (B.C.E. 570 to B.C.E. 495) was one of the first people to be known as a philosopher and mathematician. It seems that he knew all about the <u>Eternal Conflict</u> as he was obsessed with numbers, mathematics and especially music, which he considered aural mathematics. Pythagoras developed 21 central principles and the Eternal Conflict was deeply engrained within most.

Some of Pythagoras' principles was that of the cosmos or Chaos or Matter exhibited Order only because it had to obey the mathematical laws of Spirit. He believed that if the cosmos or universes made of Chaos were not mathematical, then they would have been permanently Chaotic, random and no life would have ever arisen. But Order overlapped Chaos and controls it. He also believed that God had to fill the cosmos with mathematical messages of Spirit and those hidden messages could provide all the answers to everything. Pythagoras also believed that through internal thought, the study of mathematics and philosophy is the only way to comprehend God, but it also allows you to purify your soul spiritually.

In some circles, Pythagoras is considered the first leader of a secret organization called the Illuminati, which if it really did exist today, would focus on the Eternal Conflict and the forces of Order and Chaos.

Hippocrates (B.C.E. 460 to B.C.E. 377) started what is known as the first medical schools of the Greek island of Kos. The symbol for medicine, which is the two snakes around a staff or a stick and was made the symbol for medicine by Hippocrates.

Hippocrates is considered by most people today to be the father of medicine as he developed and taught what we today call the <u>Hippocratic Oath</u>, which is the idea that physicians should heal and cure disease and practice medicine ethically. This is a seemingly a novel concept under <u>Order</u>, because the force of <u>Order</u> tends to see illness and disease to make money through the maintenance of the sickness, instead of trying to actually cure the illness or disease, which is <u>Chaos</u>.

Socrates (B.C.E. 469 to B.C.E. 399) is still considered the founder of western philosophy and the development of logical arguments and thoughts. As a philosopher, he was instrumental in the fight against the language and laws of the Phoenicians and so he helped people use language as teaching tools called pedagogy, where a series of logical questions are asked but not only to discover answers, but to also encourage fundamental insight into the discussion. His teachings about logic and language deemed him a heretic by the followers of **Order** that were still present at the time within ancient Greece.

One of Socrates best phrases is, "I only know that I know nothing", which was his way of describing the **Eternal Conflict** and how the world that is described and taught to others, is not the real world that we live in. Socrates was aware of his own ignorance and that of all humanity regarding the forces of **Order** and **Chaos**.

Plato (B.C.E. 424 to B.C.E. 347) was a student of Socrates and in his many writings describes both Socrates and his own idea that knowledge is simply a part of the God that is within us and that all knowledge can be accessed directly from God or from within us, because it comes from divine insight and not listening to other people who claim to teach you. Plato knew that learning, observation and study was important, but all knowledge was a matter of simple recollection and comes directly from the force of **Order** as the universe stores all information (past, present or future) as energy.

Both Socrates and Plato were well aware of the influence of **Order** which is invisible and therefore Plato wrote extensively about the fact that the real world is not available to those who have faith in their leaders and follow blindly. We should listen to the words of Plato even today, as today's world is mostly <u>Order</u> and invisible. The visible real world is the least known and most obscure.

In Plato's book <u>The Republic</u>, he describes Socrates as saying that, "the people who see the sunlit world of the senses to be real, are living in a den of ignorance. If you can climb out of that den of ignorance through terrible struggle, then you will see the real world and immediately desire to go back down and help other people up. However, when you do that then you will find yourself objects of scorn and ridicule to the people that you are trying to help as they will not believe you". (Plato, B.C.E. 380)

Aristotle (B.C.E. 384 to B.C.E. 322) was a student of Plato and started his schools of philosophy in Athens, Greece as place for people to learn and think about the world and the purpose of life. His Lyceum school of philosophy became the blueprint for today's colleges and were somewhat based on his mentor Plato and on the writings of Pythagoras. Again, all the schools started in ancient Greece, during that time were not public schools run by the government, but were instead individual schools that taught one-on-one, by regular people or simply a place to ponder and think.

Interestingly, science education under the ancient Greeks followed the paths of the earlier Muslim scientists, in that they tried to explain the nature world and how human body worked through observation. Doctors tried to find the causes of disease and then actually cure them, using items from nature instead of using drugs and chemicals to only treat disease.

Ancient Greece and the Eternal Conflict

Going backward in time for a moment, it is important to understand that ancient Greece started as mostly a Chaos-based society going back to almost B.C.E. 1700, during the Minoan civilization, when the Snake Goddess was being worshipped on some of the larger Greek islands, such as Crete. Also, most of the ancient Greek mythologies feature a Snake God of Chaos, such as Ophion ruling the known world until he was destroyed by the Order-God that we discussed numerous times called Kronos, around B.C.E. 1350. It is also interesting that Greek Mythology discusses Alcaeus (Heracles among the ancient

Romans) who was the son of another Order-God Zeus and who became the champion of Order against the numerous monsters of Chaos. As an infant of only eight months, Heracles was able to defend himself against Chaos by straggling two giant snakes that were sent to kill him. These multiple ancient stories reflect the Grecian culture's understanding of Chaos and Order.

It was after the Serpent Goddess and Chaos was pushed out of Greece around B.C.E. 1100 to B.C.E. 800 (the time-period that is also called the Greek Dark Ages), there was an influx of Order and a new mythology that seemed to be created around Enlil and his brother Enki, but again under different names. Enlil was then called Zeus and his father Anu was renamed Kronus or Cronus.

Also, in that early Chaos-based Minoan culture, the God Zeus was never called an all-powerful God, but he was rather considered just a mortal Demi-God that could be killed and was sometimes called Deus instead of Zeus. But later once the force of Order took over the society, Zeus became mean, arrogant and egocentric and was describes as the King and father of all the other Gods and all those other Gods had to bow down at least while they were in his presence. The same as Enlil, Zeus also kept the symbols of the eagle or birds and is frequently depicted by the Greeks, in numerous statues, as sitting in a great throne or standing but almost always with a thunderbolt in one hand.

This ancient religion described the brother of Zeus by the name of Poseidon instead of Enki and then changed his name later to Sin, the same name as one of the earlier sons of Enlil, who then changed into another brother called Hades. The most important thing about this mythology is that both Enlil and Zeus are both defined as the <u>Lord of the Sky</u> with the eagle as their symbol, while both Enki and Poseidon are defined as the <u>Lord of the Water</u> with the snake or serpent as their symbol.

Also, within this Greek Mythology is another similar story of the immortal Prometheus, who was also a Titan and who literally represented the conflict between Enlil and Enki on who was the original God to create humans from the clay of the Earth. After creating humans, Prometheus or Enki was supposedly so captivated and enthralled by his new humans that he stole fire or knowledge from the other Gods and gave it to the humans to help them with their civilization and progress. While,

Zeus or Enlil was so upset with Prometheus that he sentenced him to eternal torture and torment by chaining him to a rock and having an eagle attack and eat out his liver from within his body. Then the next day the liver would re-grow and the eagle would reappear and eat it again over and over in torment forever. This again, is almost the exact same story from the earlier Ancient Sumerians.

Zeus as the supreme God with a long flowing beard, sitting on a throne, where he will judge all humans and will deal out all punishment and torture is the archetype for the judging and vengeful God that some religions of today still have incorporated.

Ancient Greece and their Republic form of government

So, the ancient Greek affinity as a society, started with Chaos but also continued that way, when they altered their language system and rid themselves of the Phoenicians. It also came to the forefront when they created once of the first Republic form of governments, modeled after the Eternal Conflict, where most was Chaos, or freedom for the people with just a little bit of Order, which means there was some laws, but just a few.

The political experiment called the Athenian Republic, or the Grecian Republic and it started around B.C.E. 500 under the direction of politicians such as Solon, Cleisthenes and Ephialtes. Solon, as a politician, even went as far as to even create a constitution but it did not last very long. Putting this Republic in place however was not easy and was fought constantly by the followers of Order. However, with a political system setup that tried to maintain the balance between both social forces, suddenly under the efforts of the leader called Hipparchus formed the Athenian Republic, centered in Athens, Greece. If with the best of intentions however, the Athenian Republic worked well for only about one hundred and fifty years, with the best times being under the rule of Pericles and even during those times the Republic was interrupted at least twice through two different attempts at an Order-based revolution that never took hold immediately. The Republic would however eventually fail and end when the neighboring lands of Greece attacked them, the Peloponnesian Wars started and then finally the Macedonian attacks and suppression that occurred around B.C.E. 322.

The Athenian or Grecian Republic used a political system that did not elect representatives to go and vote on the behalf of the people, but instead the people themselves showed up to vote on proposed legislation or decrees themselves, and even though this slowed down the political process, it was still used because the ancient Greek people believed that a slow government was always the best government. Participation in the voting process within Athens could happen, but only if you were a citizen, which meant you were an adult male who had completed their military training. In the city of Sparta and numerous other Greek cities, both men and women could participate as citizens.

Even though political participation, as a citizen, was by no means open to everybody, it was not defined by social class or wealth and the group of participants occurred from all manner of economic class and they all participated on a very large scale. There were also three different political gatherings that Citizens could attend and vote. The first was the Assembly which needed a minimum of 6,000 citizens and these assemblies had four main functions; to make executive decrees; to make legislative laws; to elect some officials or to try any political crime. The second was the Council, which were approximately 500 citizens that also had to be over the age of thirty. The Council had one official called the President, but they only held power for one month, which prevented them from becoming addicted to either the <u>Power to Control</u> or the <u>Power of Money</u>. The common belief at that time was that any official of any government should never hold any office for very long or they would become corrupted by Power or Money. The third was the courts which needed a minimum of 200 but usually held up to 6,000 as sometimes the Assembly and the courts would blend together, and the Assembly would sometimes become a <u>Court of Judgment</u> for any trial of political importance.

All the political gatherings within the Grecian Republic usually involved discussions for and then discussions against with a direct vote right after all discussions happened. All voting by the citizens were not subject to review or prosecution. However, all council members and all officials were held to a higher standard and therefore were always subject to review and prosecution.

The Athenian Republic maintained a <u>Lawful System</u> that centered on securing and protecting freedoms, liberties and rights of the people and was regarded as expressing the will of the people. All trials were performed within gatherings and never included lawyers or judges, as they were not considered to have a valid need in the courts. Jurors were picked however from the pool of citizens within the courts. Every case would happen by first the Prosecutor and then the Defendant, in an exchange of single speeches timed by a water clock, with no more than three hours devoted to large public suits and much less time devoted to smaller suits. Decisions were then voted on immediately as to the guilt and then the sentence, without any time set aside for deliberations. So, justice was rapid and final. Most convictions triggered any automatic penalty while others involved both litigants proposing a penalty for another further vote by the jurors.

As mentioned earlier, this form of Republic did not survive long as the followers of <u>Order</u> attacked it from every angle. It is very difficult to enslave people, within a Republic, so it was slowly attacked and destroyed from within and without. For example, around B.C.E. 450 over five thousand people had their citizenship removed because it was discovered that they had been bribed by the Egyptians with free gifts of grain to support their illegal trade and banking. Also, most of the Order-based people, who are normally the wealthy constantly, wrote that a Republic form of government is terrible, because it gives the poor and uneducated people power over the rich and it gives rights to all the people instead of the elite special leaders.

It is interesting to note that during the times of the Athenian Republic, the nation of Greece prospered through education, schools of philosophy but also with major technological advances. For example, the <u>Antikythera Mechanism</u> that was discovered in a shipwreck contains very precise gears and structures that helped the Ancient Greek people keep track of time and events. Once their Republic form of government was destroyed, this type of technology will not be discovered on any technological device again, for over a thousand years.

Lastly, we are about to leave the history of ancient Greece and move into that of ancient Rome, and with that, it is interesting to note that one of the most prominent cities or states within ancient Greece was called Sparta and hence the people were called Spartans. The story goes that around B.C.E. 780, a man named Lycurgus established a very military society that was <u>Order-based</u> with very strict laws and rules. This allowed the city or state of Sparta to rise to prominence as a dominant military power within Greece, because their entire social structure revolved around military training and excellence with the military. This culture was so extremely <u>Order-based</u>, severe and controlling that a large group of them fled Greece and supposedly became the Sabines from central Italy, who would also later start the Roman Republic under <u>Chaos</u> and who would also later be suppressed by the soon to be arising Roman Empire under <u>Order</u>.

The concept called Atlantis, Lemuria or Forbidden

One last note before moving on to ancient Italy and the Roman Republic and that is that there are many different stories and interpretations of an island called Atlantis and another island called Lemuria. Both islands supposedly sunk into an ocean because of a major upheaval, which may be part of the many worldwide disasters that have occurred in the past. But, these were both probably just stories that developed to help people understand the battle that is really happening between <u>Chaos</u> and <u>Order</u>, which is the battle that common people experience between keeping their freedoms, liberties and rights or having no rights as known or unknown slaves.

The reason that I mention this, is because if there ever was such an island, then it would have been the force of Chaos that helped to create it, as according to the legend, Atlantis formed a government that consisted of ten separate Republics to help protect the freedoms, liberties and rights of the people. Then supposedly, according to the ancient writings, there was a major disagreement over how Atlantis was being managed and this disagreement split the Atlantean people into two factions, the Sons of Belial, who were Order-based angels or aliens from the star called Sirius and the other faction, called the Law of One, who were Chaos-based angels or aliens from the star called Orion. This major disagreement evolved into a civil war which led to the suppression of Chaos and the Law of One, which had attracted to the Order-based internal spiritual religions. This also caused all ten of the Atlantean Republics to dissolve and for an Oligarchy to be formed by the Sons of Belial with religions that emphasized Chaos-based materialism.

After that, both the island of Atlantis and the island of Lemuria supposedly sunk into the Atlantic Ocean never to be found again. The final message regarding both stories and that of both islands was that they had the <u>Forbidden Knowledge</u> that should never be discovered again. Again, these were probably never real islands, but rather ancient stories regarding the <u>Eternal Conflict</u>.

Chapter 13 - HISTORY OF ANCIENT ROME TO AMERICA

Etruscans and Ancient Romans

The country of Italy and the city of Rome began as a series of small civilizations that started around B.C.E. 2000 where the Latins, who were the central tribes of Italy, were able to spread their Alphabetical Language of Latin to the rest of the country of Italy.

The mythology concerning the founding of Rome and the Roman Empire is that of a woman named Rhea Silvia who was the daughter of Numitor, who was King of central Italy at that time. What happened was that Amulius, the King's brother was able to overthrow the King, kill his son and then he forced Rhea Silvia to become a priestess of the Goddess Vesta, where she took vows of celibacy. All of this was done to try and prevent King Numitor from ever having any legitimate heirs. However, later Rhea Silvia was discovered alone in the woods by the God Mars, the <u>Chaos-God</u> of War and guardian of agriculture, who then seduced her. Rhea Silvia eventually gave birth to twin boys, who were named Romulus and Remus. The <u>Order-based</u> Romulus eventually killed his <u>Chaos-based</u> brother Remus, supposedly to herald in and establish the ancient

Roman Empire, after the Roman Republic was destroyed. In reality, this story depicts the Eternal Conflict and the struggle of Order and Chaos.

Rhea Silvia as a priestess of Vesta could have been a worshipper of the earlier Vedic Religion from India, which had a comparable story of twin boys being born, the Chaos-based Nasatya and the Order-based Dasra, also known as the Ashvins or Ashwini Kumaras. These stories of twin boys being born, where one represents Order and the other Chaos goes back to the Sumerian original tale of the Gods Enlil and Enki, which we have discussed in great detail, but can also be found in the ancient Greek mythologies as Castor and Pollux of the twin Dioscuri, over in the Baltics (are of modern day Lithuania) as the twins Asvieniai, or in the ancient Hindu mythologies as Nara-Narayana. All these tales reflect both the duality of the universe and the Eternal Conflict.

This specific story of the twin boys named Romulus and Remis however, state that once Amulius learned of the twins' birth, he then sent a servant to kill them. However, in a story almost identical to the earlier Hebrew story of Mo.ses, the servant does not kill them, but instead sets them adrift on a river where a female wolf finds them and raises them as her own.

The city of Rome and the center of the Roman Empire was setup on a series of seven hills. The story continues later where supposedly, the twin boys Romulus and Remus argued over which of the seven hills to establish the Capital City on. <u>Order-based</u> Romulus wanted the main city setup on Palatine Hill, as that hill was related to the government and to the courts. In fact, the term Palatine literally meant royalty, title or class. <u>Chaos-based</u> Remus however, wanted the main city setup on Aventine Hill, as that hill was related to public lands and the common people. So, in order to settle the discussion about which hill to establish the Capital City, both of the boys performed an augury which was to determine the will of the Gods by reviewing the type and patterns of birds. In the story, Romulus wins because he saw at least twelve eagles or vultures and therefore he setup the capital city of Rome on Palatine Hill and even got to name it after his own name. His brother Remus, on the other hand, ends up getting killed by his brother, but with no further explanation or discussion, so this story probably represents what happens when the <u>Eternal Conflict</u> is out of balanced.

The surviving brother Romulus then goes ahead and sets up the city of Rome, under <u>Order</u>, because it was setup under a social order called patronage, where only the leaders could possess the wealth and power with prestige, while the followers were considered inferior, with no liberties or freedoms, but rather they must ask the leaders or the Patronus for favors or benefits. Some of the favors or benefits could simply be Legal representation in court, recommendations for priesthood or even loans of money.

The formation of the Roman Empire also centers round the date of B.C.E. 616 which was when the area of Rome was being controlled by the Etruscans or Tuscians who had arrived from the north, with the purpose of joining up with the <u>Order-based</u> Phoenicians to help support international trading and banking. Interestingly, much like the Phoenicians, not much is known about the Etruscans, other than that they also arrived with another very unique Alphabetical Language and setup Rome as an Oligarchy with their first Etruscan King named Tarquin I. The language of the Etruscans was very unique, as it had never been used anywhere else and then slowly faded away, after they adopted the Phoenician Language instead and then later morphed into the Latin Language, which became the language of the ancient Roman Empire and is still used today as the language of European and American <u>Legal Law</u> systems.

It is interesting to note how many different Alphabetical Languages have been created over time, by <u>Order</u> but these many different languages also allude back to the story of the Tower of Babel and to the Language War, discussed earlier. You can see this diversity even today, within Europe, because even though the land mass area is quite small, there are many varied and numerous languages that do nothing but separate and divide the people who live next door to each other.

Next to the Etruscans within ancient Italy was a <u>Chaos-based</u> tribe called the Sabines and they also had a separate language that would eventually disappear completely from the history of the world. It was these same Sabines that started out as <u>Order-based</u> Spartans but were the group that escaped from Sparta and the extreme laws of Lycurgus, at that time. However, just like the rest of ancient history, the <u>Order-based</u> Etruscans would work with the Phoenicians to fight against Sabines, who after losing to the larger forces, eventually became assimilated into the Roman Republic as it transitioned into the Roman Empire.

Interestingly, the time-period for the beginning of ancient Rome also corresponds directly to the flourishing of the Phoenicians and the revolution of the Grecian Republic against <u>Order</u>. Since all of these are near each other, there is probably more going on than just coincidence.

Republic forms in Rome for a little while

The actual formation of Rome occurred around B.C.E. 509, where the earlier Roman Monarchy and leadership was overthrown with help from the force of <u>Chaos</u> and was replaced by the Roman Republic. Why the sudden change in the government of Rome from an Oligarchy to a Republic is really not explained by historians, but again was probably not a coincidence. Especially since this event was probably directly related to the ancient Grecian Republic being established around the same time-period of around B.C.E. 510 to B.C.E. 500. The sudden creation of not one but two ancient Republic forms of government, near each other and of which both were setup to try and balance the forces of <u>Order</u> and <u>Chaos</u>, was overseen by the sudden appearance of a new God called Janus, who was considered by the Romans to be the God of new beginnings.

The Romans were so enthralled with this new God that they established the first month of their calendar as that of January, so that it could be dedicated to Janus. Even the word Janus is interesting as it is derived from the Greek concept of <u>Chaos</u> and is directly related to the idea that the true nature of Janus is that of a <u>Chaos-God</u>, since the force of <u>Order</u> is perfect and never-changing while the most fundamental aspects of <u>Chaos</u> are continual change, beginning, ending and beginning again. The God Janus is depicted as having two heads facing in opposite directions and of which symbolize change, progress or movement from the past to the future. In modern days, Janus is sometimes represented as Father Time, who appears as a child and then disappears every New Year's Eve as an old man.

It was also under the new <u>Chaos-God</u> Janus, ancient Rome formed a Republic that was headed by two Consuls that were directly elected by the citizens, under the advice of the Senate. The Roman Assemblies, also called the People of Rome were like those setup in Greece as they also elected the Consuls and Magistrates, who were there to create a small number of laws and to create alliances with other nations. A <u>Chaos-based</u> written Constitution were developed that was centered on keeping the main powers within the government separated and diluted. They also setup a series of checks and balances within the government and between the various powers. For example, no citizen could hold office for more than one year, which also limited the powers of any official and prevented the addiction called the <u>Power to Control</u> from happening.

Under the original Roman Republic, there was only Lawful Money that had been coined from either Copper or Brass, which was the least expensive materials and then it was spent into circulation by the government and not created as debt. Once the government provided these inexpensive coins in vast numbers, this allowed real wealth and prosperity to easily flow down to the common people and everyone thrived under this arrangement with the Roman Republic becoming one of the most prosperous nations on Earth, at that time. The importance of the idea of using cheap materials to coin Lawful Money was the understanding that it does not matter what commodity or metal backs up the coined Lawful Money, just as if it has both substance and the form of money and not based on debt. Under the Roman Republic, they also discovered that using expensive metals such as gold or silver to coin money, gave great power to the goldsmiths and silversmiths, as they could control the money supply, which in turn then also limited the amount of money that could be made and then spent into circulation which also limited the prosperity of the people.

The other thing that the Roman Republic discovered was that the most important thing about Lawful Money is that the government must create it and must also be the one to control the supply or the quantity of the money. This must never be privatized or given to citizens or private companies and it must be created without debt to the government. A man named Howard Milford wrote in C.E. 1895 that because the past Roman Republic was "without the use of either gold or silver, Rome became the mistress of commerce of the world. Her people were the bravest, the most prosperous, the most happy, for they knew no grinding poverty. Her money was issued directly to the people and her money was composed of a cheap material - copper or brass - based alone upon the faith and credit of the nation." (Howard, Milford W., C.E. 1895)

Many scientific advances also happened under the Roman Republic, including the invention of concrete, brand new construction methods, fresh water aqueducts and roads. Also, schools were formed and setup where the education of a citizen started at the age of six years and would continue for approximately seven years and where both boys and girls were expected to learn reading, writing and mathematics.

These great advancements for the common man, under the force of <u>Chaos</u>, would not last long however. The force of <u>Order</u> immediately started to advance and slowly attack both the Grecian Republic and the Roman Republic from behind the scenes and out in the open with actual wars including many self-created Civil Wars. Slowly, the banking systems and money systems would expand to fund the wars and over time both individuals and leaders soon found themselves in great debt. In addition, the many wars that happened also created the need for more and more soldiers, which sometimes forced people to leave their farms and lands to go fight for the state. When this happened, the bigger problems showed up after the people returned, as their farms and lands were now in disrepair which led to many farmers going bankrupt and losing their lands. Some of the lands had even been taken over by the rich leaders who fought to eliminate the right of people to even own their own lands and to be self-sufficient. Once the other nations, especially the Order-based nations started to refuse to even trade with those Republics, that caused the commodity prices to fall severely and suddenly many more farmers could no longer make a profit at all. This led to many more bankruptcies and then masses of unemployed and the poor citizens flooded into the city of Rome and they then started to vote for any candidate who offered them the most benefits for nothing.

It was also during this time now Order-based leaders were purposefully creating many distractions, such as national sports and events, like the Olympic style games formed in Greece, to try and fool the people of Rome into watching and paying attention to something that was not really important. The idea of Order-based distractions was to give the people of Rome something not that important to focus and concentrate on, so that the people would then not pay close attention to the fact that their country was moving away from Chaos and toward Order. Some of the distractions were created during this transitionary time-period happened in an area called the Campus Martius and this was where the Circus Flaminius was built as a large track for chariot racing plus a stadium, many temples, performance halls, circuses and monuments. It is interestingly to note that most of the Roman arenas and sporting events were <u>Chaos-based</u>, which helped to balance the ever-rising force of <u>Order</u> and this was why they were almost always dedicated to the <u>Chaos-God</u> called Mars. The largest distraction was built around C.E. 70 and was called the Colosseum of Rome and could hold up to 50,000 spectators and was mainly used for gladiatorial games and contests.

Over time, despite the Constitutional constraint against any one individual receiving permanent political power, a small number of Roman families began to take control of the entire country under <u>Order</u>. This happened mainly because of the distractions introduced into the society plus the slow increase in the preoccupation with war and conquest. Eventually, a man named Julius Caesar rose to power and he was a great military leader whose huge military power worried the Senate, so they ordered him to stand trial in Rome for various causes. Julius Caesar however, was a smart <u>Order-based</u> individual, and so he did return to Rome to stand trial, but he marched in Rome with an entire legion of dedicated soldiers. This started a Civil War from which he emerged as the new

Democratic leader of Rome, and basically ended the Roman Republic. After the Senate back down out of fear, Julius Caesar began an extensive reform of the government by centralizing the government into a Democracy and proclaimed that he was now a dictator for life and then later declared himself a God. Under Julius Caesar, all the Lawful Money that was earlier created from copper or brass was replaced with more expensive metals and larger values, so that eventually even the regular citizens could not afford to keep or use the new money being made. Also, gold and silver were the only metals being used for money and even that was no longer spent into circulation. Once, the Roman money systems were changed over to gold and silver than it only benefited the rich, the goldsmiths, silversmiths, bankers and traders. Earlier, when Rome coined Money out of cheap copper and brass and then issuing the abundant money supply, then the common people of Rome became prosperous. However, once the money supply was coined only out of the more expensive gold, then the people became poor and literally starved now as slaves.

The Senators within Rome hated the new money systems made of gold and silver. Once the new Roman Democracy got so bad, then a group of Senators rose up and assassinated Julius Caeser in B.C.E. 44 with the hopes of trying to save their country and also to try and restore the Republic. This however led to a larger Civil War, which ultimately lead to the establishment of the Roman Empire under total <u>Order</u>. The adopted son of Julius Caesar was also another powerful military leader called Octavian, later renamed Augustus Caesar and he followed his adopted father by rising up, with the loyalty of many soldiers and veterans and the fact that his control over the majority of the Roman legions created another armed threat against the Senate. This time, the Senate gave up and gave Augustus all the powers of policy, trade, money, warfare and diplomacy and allowed him to hold these powers for his lifetime.

Augustus Caesar went even as far as to remove any of the older money that was coined from Copper and Brass from circulation and then issued coins that only had higher value amounts. This happened because the bankers effectively reduced the supply of money by over 90% and created a massive depression on purpose. All the wealth of the people was quickly moved from the common people to the wealthy bankers and traders, which is one of the effective ways that <u>Order-based</u> people can use to destroy any Republic form of government from the inside out. It is a fact that whenever any government purposefully starts to destroy its own money systems, then it is just a matter of time when it will also fail politically and socially which will turn it from a Republic to a Democracy and eventually to an Oligarchy. As mentioned earlier, Cicero was a famous orator and philosopher from Rome and one of his most famous quotations is actually a summation of his political philosophy, it can still be used as an accurate warning that Cicero gave to Rome about it destruction of its money supply by the <u>Order-people</u>, when he stated that "the budget should be balanced, the treasury should be refilled, public debts should be reduced, the arrogance of officialdom should be tempered and controlled and the assistance to foreign lands should be curtailed lest Rome become bankrupt. People must again learn to work instead of living on public assistance". (Caldwell, (Janet) Taylor, C.E. 1965)

Under Julius Caesar, the Roman Republic started its transition to a Democracy which would under Augustus Caesar become an Oligarchy. In B.C.E. 27 after the Roman Senate granted of all the extraordinary powers to the now called Augustus Caesar, who was also declared a God, the Roman Republic was gone forever. This was about the same time-period that the Grecian Republic would also fail.

Note that the importance of both Julius Caesar and Augustus Caesar as dictators and officially ordained Gods cannot be underscored. In fact, the Roman calendar of today originally had ten months with October (Octav) being the original eighth month, November (Nona) being the original ninth month and December (Decim) being the original tenth month. However, after the two Caesars were declared to be Gods which led to the destruction of the Roman Republic, they were each given two extra months, July for Julius Caesar and August for Augustus Caesar. Suddenly the Roman calendar now had twelve months and the term Octav, which originally meant eight, then becomes the tenth month; the term Nona, which originally meant nine, then becomes the eleventh month; and Decim which originally meant ten, then becomes the twelfth month. Through <u>Order</u> comes <u>Chaos</u> and confusion.

Throughout history, you will discover the same pattern of the force of <u>Chaos</u> establishing a Republic politically, but then being out-of-balance, the force of <u>Order</u> will make sure that it never lasts more than 100 years, with the average being much shorter. It is also interestingly to note that once the Roman Republic was destroyed, the new symbol or standard for the Roman Empire, which was also used as the standard for the Roman

Legion Army, was called the Aquila and was formed in the shape of an eagle representing the advent of <u>Order</u>.

<u>Extensive Legal System setup in Rome and still in use today</u>

Around B.C.E. 439, when Rome was still under a Republic form of government, the system of law for the people was a very simple one and was called the Twelve Tablets. The Twelve Tablets formed the centerpiece of the Roman Constitution and was just a list of basic Freedoms, Liberties and Rights, with the addition of established procedural rights for all Roman Citizens. They were inscribed on twelve ivory or bronze tablets which were posted in the government area called the Roman Forum for all Roman Citizens to read and understand. The Twelve Tablets came about when a group of appointed Romans went to Greece to study their new Constitution and their Republic form of government.

However, once the Roman Republic was eliminated, then the simple system of limited laws was replaced with extensive and complex Legal Laws. The destruction of the Twelve Tablets was so complete that the original text of the Twelve Tablets has even been completely lost from history and the supposed remnants of the lost text that were found are so strange that they cannot be believed and are probably forgeries.

By the year C.E. 529, a new extensive Legal Law system came out of nowhere that was setup under Order and was called the Corpus Juris Civilis or the Justinian Code. This extensive Legal Law system was started under the Roman Emperor Justinian I and was effectively used all the way until C.E. 1453 - during which it also influenced the Canon Law of the organized Christian Universal Church, at that time called the Roman Catholic Church.

There were four parts to the Corpus Juris Civilis, with the first part being called the Code or Codex and was simply a huge compilation of all the Roman Imperial decisions that had been made to date. Further decisions were then added as they happened. The second part was called the Digesta or Briefs and was a huge encyclopedia composed of all the writings from all the jury decisions, sometimes also called Brief Extracts, hence the term Legal Briefs still used in even today's Order-based Legal Systems. The third part was called the Institutes and was a huge textbook that introduced all three parts and explained all three in conceptual terms for use in Legal schools.

The enactment of this new Legal Law system happened so quickly that the first printings of the Corpus Juris Civilis only stated three sections. These were given the force of Legal Law and were intended to be the only sole source of all Law. In fact, once this system was setup - any reference to any other source of Law was forbidden.

Later, as additional laws were put in place under Emperor Justinian I, a fourth section was added that was then called the Novels or New Laws. The text of the original Corpus Juris Civilis was composed almost entirely in Latin, but it was later also composed in Greek for use in International Trade with merchants, traders, sailors and bankers. This Order-based Legal System was so effective that most of the new Legalize Language from Black's law dictionaries still uses the original Latin phrases that come from the original Corpus Juris Civilis - phrases such as In Propria Persona meaning to act without a lawyer, or Quid Pro Quo meaning a fair exchange or getting something in exchange for something else. The four parts of the Corpus Juris Civilis today constitute the foundation documents for all the western Legal System and it continues to have great influence on the International Trade or Commerce Laws of today.

Ancient Macedonia and the world war started under Order

In B.C.E. 338, as the Grecian Republic was being destroyed, the country of Greece was invaded and came under the rule of a Macedonian King called Philip II and his son Alexander. <u>Order-based</u> Alexander who was also called <u>Alexander the Great</u> went on to become the greatest leader, with the greatest army of all time and he built one of the largest empires that stretched throughout Europe and from Africa all the way to Asia.

There are several legends regarding the birth of Alexander and one says that a thunderbolt from the Order-God called Zeus struck his Mother called Olympias in her womb on the eve of her marriage to Philip II and thus she became pregnant with the Demi-God Alexander, as his real father was Zeus. The story goes that Philip II was still supportive of Zeus, even after this the birth of Alexander, to the extent that he was helping Zeus to eliminate both the Grecian and Roman Republics through attacks and wars. Supposedly however, as Alexander got older and even though he was the son of Zeus, he didn't support the wars that Philip II waged for Zeus against Rome and Greece. The main reason was that Alexander was born in the city of Pella, Greece and was supposedly taught directly by the famous Greek philosopher Aristotle, so he would understand both the forces of Chaos and Order, as well as the Eternal Conflict. It was however, after the arrest and death of his Greek teacher Aristotle, that Alexander the Great suddenly rose up, out of fear and with the complete ambitions of Order to become the world's greatest general of the greatest army of his time.

Interestingly, Alexander the Great and his armies traveled around the known world and specifically attacked all the Chaos-based cultures and nations. It was as though, he understood the nations that had been built upon Chaos and he attacked and was undefeated in all battles against those Chaos-based cultures. Once Greece was no longer a Republic, he then led a coalition of Greek states and went to war against Persia around B.C.E. 336 to stop the rise in the religion of the Chaos-based Zoroastrianism and in just two years, he conquered the entire Persian Empire. He then led a coalition of soldiers against India and the rising Hindus of the Indus

River Valley. After that he went up into China to try and stop the rise in Confucianism and Taoism there. By the age of thirty, Alexander the Great had created one of the largest empires of the ancient world, as it extended from the Ionian Sea all the way to the Himalayan Mountains in China.

Then something weird happened, as around B.C.E. 324, Alexander the Great turned against Order-based nations, as he moved his armies into the Middle East, with the plans to invade Mesopotamia and Arabia, which at the time were all Order-based nations, still under the worship of the Order-God Enlil. Why he suddenly changed his mind and attacked Order-based nations will probably be never known but the result became what is described as the Mystery of the Death of Alexander the Great, in the palace of Order-based King Nebuchadnezzar II of Babylon.

It was as if Order was alive during those days, because ironically by the year B.C.E. 168, the new Order-based Roman Empire attacked and conquered both the lands of Macedonia and Greece, which were the original homes of Alexander the Great. The Roman Empire then went on to do what Alexander only dreamed of, which was to conquer most of the known world.

Where did all the Gods go?

The fall of the Roman Republic and the Greek Republic is interesting for many reasons, but one of the strange things that also occurred during that time period, is that for the last time in recorded history, both the <u>Chaos-God</u> and the <u>Order-God</u> were seen among the people of Earth. Suddenly all the Gods disappeared, both within the stories and mythologies of nations, but also within any eye witness testimony of the people. No longer would there be any mention of the Gods actively working with people and nations. Suddenly, there is no longer any mention of the sons and daughters of Gods walking among the people. It is as if the leaders of the modern world suddenly realized that with all the new <u>Order-based</u> systems, there was no longer the need for such things and once that change took place, then a new calendar was created from within Rome, to mark the occasion. This calendar was to be used by the entire world and was called the Gregorian calendar. It is still in use today.

The Gregorian calendar divides the time frame of the world into a <u>Before Common Era</u> period marked with

B.C.E. which is also the time periods where the Gods walked the Earth among the people, and also a later period of time called the <u>Common Era</u> period marked with a C.E., which is when the Gods suddenly disappeared from all mythologies, stories and writings.

I would also add that in my opinion, it was during this same time period that a new phrase was created in the Rome Empire and this phrase still exists today and it is <u>Novus Ordo Secorum</u> or the <u>New World Order</u> and this is the phrase that is even printed on the U.S. Dollar bills that are used as paper currency for the world, even today. If the <u>New World Order</u> is the <u>Common Era</u>, then I would guess that <u>Before Common Era</u> must have been the <u>Old World Chaos</u> - at least to the world leaders of today.

<u>The prophet Muhammad and the fight against Order</u>

During the rise of the religion and government system, called the Roman Catholic Church, started within Italy and expanding eventually through all of Europe, there was a major trading system that was happening in the Middle East and more specifically the newly called Arabia, which was to the west of Sumer or modern-day Iraq. During that time, Arabia was a very difficult land to live in due to hot climate and frequent sand storms, but the traders and bankers within Arabia controlled what was called the Frankincense spice and so certain regions of Arabia became very wealthy.

This Frankincense spice was used within most of the <u>Order-based</u> religions of that day and therefore was used in almost every religion throughout the Middle East all the way up and into Europe. The Order-based trading and banking groups and their trade routes extended from their home base in the city of Mecca, within Arabia, all the way up to the Mediterranean Sea. The city of Mecca, at that time was the home base for many of the bankers and traders that worked those routes. The Arabia Peninsula at that time was very <u>Order-based</u>, and so it kept most of the common people with little or no education and so they worshipped multiple Order-based Gods and Goddesses based on the sky and the air. During that time, it is said that just the city of Mecca alone had almost four hundred different statues of different tribal Deities.

It was within this culture that around C.E. 570, the prophet Muhammad was born in the city of Mecca. But at an early age, he became an orphan and Muhammad was then sent to become one of the traders within the <u>Bedouin</u>

tribes. At a very early age, Muhammad adopted the practice of meditating alone or searching within himself for the Spirit side of God. Whether it was due to the traumatic death of his parents or some other reason, Muhammad would go and meditate "with his family for a month of every year to a cave in the desert for meditation. His place of retreat was Hirâ, a desert hill not far from Mecca, and his chosen month was Ramadân, the month of heat". ('Ali, 'Abdullah Yusuf, M.M. Pickthall, C.E. 1999)

It was during one of these times that an angel appeared to him in the year C.E. 610. This initial first revelation was followed, after a pause of almost three years during which Muhammad searched deeper within himself, looking for God and he used many more spiritual practices besides just meditation. Sometimes the meditation and the knowledge he gained from it affected him physically and even caused him minor seizures.

Even though Muhammad originally worked as a trader, within the actual banking and money systems, it was at the end of his third year of waiting that he received a command to "arise and warn, whereupon he began to preach in public, pointing out the wretched folly of idolatry in the face of the tremendous laws of day and night, of life and death, of growth and decay". It was soon after this that Muhammad overtly rejected the trading and banking systems and it is important to note that when he was preaching and warning the people, he was speaking about the duality of the universe and the <u>Eternal Conflict</u>. ('Ali, 'Abdullah Yusuf, M.M. Pickthall, C.E. 1999)

Muhammad even started preaching against the love of money and trading, which only takes people's wealth and makes them poor to enslave them. He did this even though trading and banking was a significant portion of the commercial life of Mecca, at that time. Later, Muhammad began to develop a following among the people and that made him a threat to the some of the local tribes of traders and especially the leaders of the cities, whose wealth was created by stealing it from the people, and of which Muhammad's preaching was threatening to overthrow. Some of the most powerful merchants and traders even tried to convince him to abandon his preaching and even went as far as offering him admission into the inner circle of merchants but Muhammad could not be bought off with money, so he simply refused.

By the year C.E. 622, there were many assassination plans being put together by the Order-based traders and bankers, within Mecca to kill him, so he escaped to the city of Medina. However, within Medina, there were many religious problems with Hebrews and Christians, who were living with the local people that also worshipped numerous other Gods. So, one of the first things that Muhammad did was to teach the acceptance of other religions and to draft a document called the Constitution of Medina that went a long way toward settling most of the long-standing grievances that all the different groups had. This Chaos-based approach that Muhammad used by accepting other people and other people's chaotic viewpoints, helped to establish the alliance of eight specific tribes, which then secured a series of freedoms, liberties and rights for all people.

Muhammad first tried to create a non-violent revolution against the traders and bankers and that at least seemed at work in the beginning. However, the Order-based traders and bankers could not afford to let Muhammad and his new religion change the country, so they used the money and trading systems to cause suffering and problems with all the people. They caused so many problems that Muhammad finally had to turn against them violently with actual war. This physical confrontation showed the entire country of Arabia the truth about the real economic system that was setup against the common person, so suddenly they developed a thirst for knowledge and turned away from the Order-based systems.

One of the most important things to understand about Muhammad was that to him, the true religion was the fight against slavery, not just as an individual but as a society that fights for economic, social and political justice. I believe that Muhammad saw Islam as a religion but also as a type of political revolution against any Order-based tyranny, that occurs once Order becomes too strong and there is no more balance. Once Muhammad understood the Eternal Conflict and the two forces at work within the universe, I believe that he proclaimed himself to be the Final Prophet because he knew that history just repeats itself over and over again, at least until a balance can be achieved between the force of Order and the force of Chaos.

I mentioned earlier that today's world leaders are mostly Order-based and so the information has been flipped over. Instead of fighting against the Order-based trading, money and banking systems, which Muhammad rallied against, today the Muslim world has been tricked into fighting against Christians, Hebrews and all other religions, even though most of those religions worship the same God of Abraham, Issac and Jacob and even though Muslims have always even accepted Jesus as the perfect prophet who never sinned. Especially since, Muhammad himself helped establish the Constitution of Medina that brings understanding of the chaotic beliefs of other people.

Rise of Muslim scientists

Between the years C.E. 600 and C.E. 1400, which is when the followers of Muhammad were able to galvanize the entire country of Arabia against Order-based systems, by uniting all the tribes of Arabia under the guidance of Muhammad, it was only then that Arabia was able to take over all of the Middle East. This was also the time-period where Order began to retreat from the Middle East and move instead into Europe, where the force of Order went out-of-balance and was too soon cause the European Dark Ages, where knowledge would become suppressed and severely controlled.

Once the force of Chaos moved back into the Middle East and balance started to become normalized again, the

Arabian people suddenly flourished and their knowledge greatly expanded. The Arabic scholars were even able to translate the ancient Greek philosophical writings and Hindu writings into the new language of Arabic. Muhammad was not able to even read when he began preaching to his people, but he did instruct all of them to read and gather knowledge. Interestingly, that also included even the sacred writings of all the other religions, plus he fought against Censorship of any kind. It was Muhammad himself that demanded that his people stay open minded and study just for the sake of learning.

Muhammad was proven correct, because one the availability of uncensored knowledge occurred during these days, it helped to create a scientific revolution with the Islamic world. For example, although <u>Geometric Algebra</u> was invented in ancient Greece (as a simple relationship between shapes and mathematics), what we call today as <u>True Algebra</u> was actually invented around C.E. 800 by an Islamic Scientist named Al Khwarizmi, who created a new kind of mathematics that he called Al Jabr or what we today call Algebra, and which is completely independent of either geometry or arithmetic. The uncensored knowledge of those days allowed one of the greatest advances in pure science that came from a Muslim scientist who learned to place greater emphasis on __experiment__ and __observation__ than anybody else.

It was around the year C.E. 1000 that the great Muslim scholar called Al Haytham created the modern scientific methodology with experiments and observation. He wrote over two hundred books and revolutionized science, physics, optics and momentum. He also created the first camera and the first telescope and explained correctly how the eye actually worked.

Arabia, as a __Chaos-based__ country, actually became a shining example to the entire world and would eventually change its name to Saudi Arabia, once the commodity called oil was discovered in the Middle East.

The Order-led Chaos-followed Church of Rome controls Europe

Around the year C.E. 600, under the guidance and writings of the final prophet Muhammad, the Muslim people stopped fighting each other and started a worldwide conquest against __Order__. It was also during this time that the so called __Dark Ages__ of Europe began.

Around the year C.E. 650, the Muslim armies were able to take back all the lands east of the Mediterranean Sea, including the city of Jerusalem from the Roman Leaders, who were now called the Byzantine Emperors. The Muslim armies at the time did not exile either the Jews or the Christians from even the city of Jerusalem, but instead just made them pay a special charge or tax to stay in the city.

Around the year C.E. 800, when the Order-based leader named Charlemagne founded the Holy Roman Empire by going to war with most of Europe and literally forcing them to accept organized Christianity under the control of the Roman Catholic Church or be tortured or be put to death. Once this new Holy Roman Empire was created, the new standard of the Holy Roman Empire changed to become the eagle representing Order and the Order-God Enlil.

Once that happened, then the normal Hierarchy of Laws was inverted by the Roman Catholic Church, whereas the first of the laws, God's Law was changed. Normally, the Laws of God are given to everyone as freedoms, liberties and rights, but the Roman Catholic Church under the new Holy Roman Empire changed it, by announcing that only one man, the leader of the Roman Catholic Church called the Pope was the only one that got freedoms, liberties and rights directly from God. The Pope then could give those powers to certain <u>Order-based</u> Kings or Queens, through a coronation ceremony. Those powers were, of course, called the <u>Divine Right to Rule</u> and the government that was setup was one where the common people never got rights only privileges and only if they qualified.

Around the C.E. 1009, while the force of <u>Order</u>, under the <u>Roman Catholic Church</u> was continuing to achieve total control over the Kings and Queens of Europe, the Muslim armies aligned with the force of <u>Chaos</u> once again took back all the lands east of the Mediterranean Sea, including the city of Jerusalem.

Around the year C.E. 1184, the <u>Roman Catholic Church</u>, who was basically the dominant political party in Europe at that time, introduced what would later be called the Medieval Inquisition or simply the Inquisition. The Inquisition was a Legal System developed by the <u>Roman Catholic Church</u> to censor any writings or speech, especially heresy using horrendous torture and viscous methods to induce pain. What happened was that individuals were brought before the Grand Inquisitor and read the charges of their guilt of heresy and then they were asked to admit their guilt. The process of Inquisition never allowed a defense or a trial but simply required an admission of guilt by the accused to the Grand Inquisitor who was Order-based and always correct. If the individual admitted that they were guilty, then they would be tortured for their crime. However, if the individual refused to admit their guilt then they would still be tortured until they admitted their guilt and then after would be given to the government officials to be burned at the stake or given death by fire.

This is the time-period in history called the European Dark Ages, where the Roman Catholic Church repressed all education, including reading and writing while most unapproved books and knowledge was destroyed. There was also a scarcity of created art or literature during those times.

It was during the rise of the Roman Catholic Church, as more of a political organization then religious, that many of the complex cathedrals were built, as once again when Order reigns, complex architecture is created. These constructions were also usually financed by the group called Knights Templar, which were, at the time, the finance arm of the Roman Catholic Church. In fact, most of the most elaborate and complex cathedrals in Europe were built by the Knights Templar working with various stone masons.

The European Dark Ages ended with a movement that led people away from the Order-based leaders of the Roman Catholic Church and the creation of numerous other Christian groups, such as the Eastern Orthodox Church and the reformation movements, such as the one started by Martin Luther. In reality, the European Dark Ages, was the specific period of time between the creation of the Roman Catholic Church by their Order-based leaders and it ended with the resurgence of Chaos at the start of the Renaissance Period, where questions and inspiration were finally allowed to flourish, censorship was stopped, and this led to the birth of new ideas, art and literature.

Brotherhood of the Snake continues based on Chaos

Out of all the animals that you can find on Earth, there was none was as noticeable or as sacred as the snake or serpent throughout our ancient history. As mentioned earlier, there was a secret society established called the Brotherhood of the Snake to help people learn sacred knowledge, truth and freedom, which are all based around Chaos.

Since the expansion of the Brotherhood of the Snake society into the world, the ancient stories of the serpent races had also expanded. These stories talk about certain Gods that have lizard-like or serpent-like features that want to help humanity by increasing knowledge and promoting justice, freedom and liberty. These stories have circled the globe and include the Dragon-Gods in Japan, the Serpent-Queen called Nukua in China, the Reptilian-Gods in Australia and the Naga-Serpent-Gods in India. There are also specific stories regarding Serpent Gods that supposedly created humans, from the ancient Mexican Mayans and Aztecs, the Hopi Indians of North America, as well as Adohwedo from Africa. There are also two specific North American Indian tribes called the Iroquois, whose name means serpent and the Sioux, whose name means snake.

It is interesting to note that since the days that the Brotherhood of the Snake expanded into the world, everything is backwards, because today everyone is taught that the snake is evil and the worse animal ever. Biblical we are even taught that the snake actually caused humans to sin for the first time within the corrupted story of the gardens in Eden.

Native Americans in North America and Canada

World History would not be complete without mentioning the Native Americans that are today incorrectly called the Indians of North America and Canada. Most of their history, at least that which is taught, is simply that they were killed, and their native lands stolen by the European settlers that started arriving to the Americas around C.E. 1500. Their real history is much more colorful and wonderful, especially regarding the Eternal Conflict and the struggles that they experienced as Chaos-related tribes and people.

As has been mentioned, the forces of Chaos and Order seek to achieve balance and will do as natural forces. While the Order-Gods were making war within Egypt, the Middle East and Europe and Order was growing stronger in those areas, other areas of the world had Chaos growing strong. Those areas, such as North America, Central America and South America were pretty much based on Chaos. This is the major reason why so many of the Native American tribes that existed during those same times took the names of snakes and serpents and build societies based around choice and freedom.

Most people have heard of the North Americans earthen mounds, especially those found in Midwest states such as Ohio, but what most people may not know is that when they were built they were built as large pyramidal structures made of the Earth and dirt. Most of them have settled and collapsed over time and are currently referred to as simply mounds, but under the force of <u>Chaos</u>, where built as earthen pyramids. The largest of these smaller types of pyramids was called the Monks Mound in the State of Illinois and it is over 100 feet high with a base that runs approximately 955 feet by 775 feet. If they weren't built in the shape of a pyramid, then they were built in the shape of a snake, such as the famous Serpent Mound found in Pebbles, Ohio. The Serpent Mound is a 3-dimensional snake mound, built almost a quarter mile long and which represents an uncoiling snake. It was constructed around C.E. 1000.

Regarding the effects of <u>Chaos</u> on the Native Americans, most of their tribes used Logographical Languages, with the most notable being the Mikmaq Hieroglyphs used by the Mikmaq tribes and the

Wiigwaasabak Language of the Ojibwe tribes, who were both located in Canada and the northern areas of the United States of America. Most of the early Native Americans also had <u>Chaos-based</u> governments that were very much like Republic forms of governments, whereas each tribe had the right to elect their own chief or leader and elect each of the members of the Council of Chiefs, which acted very much like a Senate. All individual tribe members had inheritance rights of their deceased family members, obligations to help and defend all the members, the right to give names, to adopt strangers into the tribe and common religious rights.

Speaking of the religions within the Native American tribes, they were all grounded in the fact that there is a Wakan Tanka or a divine essence within everything, which is like saying that God is in everything and that everything is in God. It is also interesting to note that almost all the Native American tribes have the same mythological story which is called the <u>Thunderbird and Trickster</u>, which speaks directly to the conflict between <u>Order</u> and <u>Chaos</u> within all their varying religions. This story relates the thunderbird to an <u>Order-God</u> like Enlil, who is a huge monster that must be slain by a hero. The thunderbird is the personification of the ever-present energy that is found in nature and that can generate thunderstorms or thunderbolts.

In many Native American tribes, the creature called the thunderbird was also called the Wakinyan or the great eagle that produces thunder from its wings and flashes lightning from its eyes. Some legends also say that the thunderbird is malevolent and carries off humans to their death and enslaves people for no reason. It is interesting to note that the thunderbird in most Native American mythologies is not considered a natural physical bird but would be better described as a spiritual essence that brings destruction as well as <u>Order</u>.

Linked together, but the opposite side of this mythological story is that of the trickster or clown and this can also be found in almost every Native American culture. The trickster is not spiritual at all but rather purely physical and is related to the <u>Chaos</u> or the craziness of nature. The behavior of the trickster, who is like a <u>Chaos-God</u> such as Enki, is to help humans by joking around, causing laughter, playing practical jokes and of course taunting the enemies of the people. The trickster carries out expected performances during almost all religious ceremonies, such as Crazy dances or riding a horse backwards. This is done to represent its Chaotic nature.

Interestingly, the Ojibwe tribes often wore masks which made them appear to have two faces, like the earlier described Chaos-God called Janus of the ancient Romans. The masks are worn to symbolize Chaos or the freedom to do something weird that was normally forbidden. So, indirectly, the trickster is the Chaos part of life that desires freedoms, liberties and rights and the ability to decide your own fate and make your own decisions.

Within some of the Native American tribes is also the story of a white bearded God that was originally a lizard or a snake but that came to help each of the Native American tribes.

The American Colonies of England

There are literally thousands of examples of the Eternal Conflict within our world history, such as when both Napolean or Hitler moved their nations forward toward pure Order, why did they remove the old symbols of their country and then install the standard of an eagle in their place? But for this current discussion, we will end it with the founding of America as a nation. Mainly because its history, from the founding of the American Republic to its current situation is almost identical to the earlier Grecian and Roman Republics, from our past. The story of America occurred with the years C.E. 1754 and C.E. 1763, where a series of wars between the French colonies of Canada attacked down and into the British colonies of America. These were incorrectly called the <u>French and Indian Wars</u>.

The importance of these conflicts was that England or Britain shipped over troops to fight for the American colonies against the invading French troops, who eventually lost the war. Keep in mind that the American people, at that time were still a colony of England controlled through the English Oligarchy under King George III. Most of the <u>Order-based</u> control that the American colonies experienced happened because of the English trading systems and English tax systems, setup under the <u>East India Trading Company</u> using English Money. The American colonies were forced to speak the English Language systems using both uppercase and lowercase letters, even though the people that made up the American colonies were originally from many different lands and spoke many different languages. The Legal System and banking systems were also from England, even though British government had a difficult time trying to control the money supply of the American colonies. This was mainly since the British Pound was a true weight in England but varied in weight, by the time that the money supply physically got to the American colonies, because the traders would literally shave off the edges of the silver coinage and steal some of the metal. In fact, the Colonial Pounds varied so much that the colonists usually relied on the Spanish Dollar instead, which is the actual reason that even today the American money systems are based around the term dollars.

<u>Order-based</u> England was very upset about the lack of monetary control over their American colonies, that they outlawed the American colonies from coining, minting or even printing its own money - which would have directly benefited the people within the colonies. The money problem got larger however, when the Massachusetts colony went ahead and built a mint and started coining its own silver money from the year C.E. 1652 to C.E. 1683 and then started printing its own paper money in the year C.E. 1690 and called it Colonial Script.

After the French were defeated after the <u>French and Indian Wars</u>, England then demanded that if the American colonies were going to print their own money, than it had to be Legal Tender money and it could only be used for public debts and it could never be used to pay for trading, which at the time was private debts or money that was sent back to England. In fact, in the year C.E. 1764, England under the orders of their Central Bank called the BANK OF ENGLAND demanded that the American colonies immediately stop the issuance of Colonial Script. At the same time, the British government under control the BANK OF ENGLAND started taxing the American colonies, supposedly to pay for their past war efforts in the earlier <u>French and Indian Wars</u>. The new taxes were just a way to punish the colonies for trying to use a different money system and this created massive tension between the American colonies and England, especially since these taxes became increasing large and burdensome to the point that even the British citizens that were living in the American Colonies started to hate them. In other words, the force of <u>Order</u> became too large and out-of-balance.

The American colonies then formed what they called the First Continental Congress and this group sent letters to King George III asking him to stop the Coercive Acts, which involved money, taxes and trade. They never got a response, so then the new revolutionists finally made the decision to go to war. In the month of May in the year C.E. 1775, a new government was formed called the Second Continental Congress, which declared the start of the American Revolutionary War after the battles of Lexington and Concord. The new American Congress voted to create their Continental Army and appointed George Washington as the commanding General of the new American army. They also appointed the first President of the country and his name was Peyton Randolph. Congress then ratified the Articles of Confederation and sent their Declaration of Independence to King George III.

The result was the <u>American Experiment</u> called the War for Independence, which started out as a battle against <u>Order</u> in the form of money, taxes and trading but ended up being a battle for <u>Chaos</u> with the founding of the American Republic and individual freedoms, liberties and rights, which became the true American Dream. The reason that we know that America was founded based on <u>Chaos</u> was that once the Declaration of Independence was issued to England, the first American flag was created and although it could have been created with any symbol or design, it was created with a large snake, representing the Chaos-God, placed onto a field of yellow with the words "Don't Tread on Me" imposed underneath the snake.

The goal of the American Revolutionary War started out as a battle against money and taxes, but it then turned into a reconstitution of the Hierarchy of Laws away from the perversion created by the earlier Roman Catholic Church whereas the first of the laws, Gods Law or the Laws of God would be revised back to their normal order. If you remember, during the start of the Roman Catholic Church that the Laws of God were incorrectly flipped over and the freedoms, liberties and rights of the people that were inherent rights were instead taken away and given to the Pope and then Kings and Queens through the coronation ceremonies. The new country called the United States of America corrected this problem immediately, by flipping the perversion back over to the correct order. Their Declaration of Independence states that all men (people) were created equal and endowed by their Creator with certain una-lien-able and inherent rights. That first sentence reconstituted God Law as the first of all laws under the Hierarchy of Laws and reestablished that all people get their freedoms, liberties and rights directly from God, so no government can ever violate them. It also established the original American Dream, which was one of establishing a Republic form of government that was extremely limited and one where all the people had freedoms, liberties and rights that could never be taken away by the government.

The American Experiment comes under attack immediately

The American Revolutionary War is a very strange war in its details, whereas some battles are not finished while others are not even started. Remember that most of both the American and British troops were all friends that had just fought together earlier on the same side during the French and Indian Wars, so I must assume that they might have had trouble fighting against each other now. For example, the British armies easily invaded and captured New York City and made that city its base of operations.

Also, most of the battles that George Washington fought were ones that he lost, so he had to retreat from them numerous times. Eventually his army got trapped in an area called Valley Forge, where they spent six months trapped in the middle of winter, freezing and starving while also being ravaged with sickness and disease the entire time. The weird part was that during this entire six months, the English armies never attacked to finish them off and at the end of six months, his army was back in wonderful shape and still able to cross the Delaware River and win their first major battle, which makes no sense at all, especially if the troops trapped at Valley Forge never had adequate water, food or even enough clothing for six months. An interesting note to this story is that during the time where they were trapped at Valley Forge, George Washington was supposedly confronted by a female angel who appeared to him and gave him a <u>series of visions</u> about all the coming wars.

Even given all the strangeness of the American Revolutionary War, we are taught that America fought the war for Independence from England, but you are almost never taught that America never won the war, but instead a truce was negotiated by Alexander Hamilton and that of course the truce involved money, banking and trading, which was the very reason that the war was being fought in the first place. America as a colony was fighting against the <u>Order-based</u> money systems of England and because of that, all during the American Revolutionary War, there were other countries, such as France, Spain and the Dutch Republic that were working with America and against <u>Order</u>. Most of these countries also supplied ammunition and weapons to the colonists. France also went as far as threatening an invasion of Britain directly and Spain helped by physically removing the British soldiers from the now state of Florida. In the end however, America agreed once more to accept the Legal money, banking and trading systems of England as well as the Legal laws and Court systems of Britain. In exchange, a truce was called and neither America not Britain won.

You may be taught in schools that the British armies surrendered in the city of Yorktown in the year C.E. 1781 but this is just a lie as the American Revolutionary War continued to be fought for another entire year, before the real truce was called. The fact that America did not officially win the <u>War for Independence</u> is why the American colonies did not take any war spoils. In fact, after the truce was called, the British General named Charles Cornwallis was even allowed to take his 8,000 troops, all his armaments, artillery and ammunition back to England without any discussion. Mysteriously, one of the first acts of the new American

Congress was to redeem all the now worthless paper money that had been printed in America at 100% of its face value, by exchanging it for coined silver money, of course obtained from the BANK OF ENGLAND.

In the year C.E. 1783, the peace Treaty of Paris was signed, and the lands of North America were separated between England, France and the newly formed United States of America. Interestingly, the first line in the Treaty of Paris acknowledges the truce, as America had to acknowledge that King George III was by the grace of God still the King of Great Britain, France, and Ireland plus being the defender of the faith, duke of Brunswick and Lunebourg, Arch Treasurer and Prince Elector of the Holy Roman Empire. This directly implies a solid connection between the Church of England and the Roman Catholic Church which we are all taught had been completely separated by that time. The truth is that a truce was called, and the American Revolutionary War was stopped after a successful negotiation by Alexander Hamilton with the country of England. Within the negotiations, the American colonies promised to still acknowledge the sovereignty of the King of England and to still acknowledge his Divine Right to Rule. Most importantly, America had to immediately setup a Central Bank within the new United States of America that was to be chartered and controlled through the BANK OF ENGLAND and the CITY OF LONDON CORPORATION which are both directly connected to the HOLY SEE and the Roman Catholic Church, even to this day.

During this time, Alexander Hamilton became very wealthy because of his secret negotiations and the establishment of the first Central Bank inside the United States of America. This created a rivalry and tension between Thomas Jefferson and Alexander Hamilton that never ceased. The first Central Bank was called the BANK OF NORTH AMERICA and was setup in C.E. 1781, modeled and chartered after the BANK OF ENGLAND and it existed for only three years before the people revolted against it. The Bank had literally forced the new United States of America to go into so much debt, in such a short time that inflation became rampant. However, that didn't last long as England would never allow America to become free of their money, banking and trade control for too long. This was around the year C.E. 1787 and was when America finally got around to writing and approving their Constitution for the United States and these late dates are why the earlier first documents, such as their Declaration of Independence written in C.E. 1776 always has more force regarding the inherent freedoms, liberties and rights of the people based on Gods Law and not the later Constitution.

Later, the second Central Bank occurred in the year C.E. 1791 and it was still called the <u>First Bank of the United States</u> again with a name to deceive the people as the Central Bank was again privately owned and not a part of the government at all. This one lasted for twenty years until C.E. 1811 before it was again removed and this time the BANK OF ENGLAND did not wait long before they started the <u>War of 1812</u> against America for once again not letting England control their money supply as promised, in the unwritten negotiations that created the truce to end the American Revolutionary War.

In reality, the American colonies were given a specific land division which was to become the United States of America if they would continue to be a British colony behind the scenes. Therefore, the country of America today charters their money systems, banking systems and even legal systems through England as they are a secret colony of England. If you doubt this, understand that the Legal System in America is still the same Legal System that was started in the early days of the Roman Empire but is still chartered through England under the <u>British Accreditation Registry</u> or B.A.R Association, which all American judges and American lawyers must still follow. This is also why the first flag of the United States of America, established under <u>Chaos</u> with a yellow background and a snake on it was immediately eliminated and a new flag was commissioned that was identical in design to that of the British <u>East India Trading Company</u>. The earliest symbol of America was the snake but was also replaced with that of the eagle, representing <u>Order</u> and this symbol of <u>Order</u> as an eagle continued to the moment when America supposedly made their infamous Moon landing and to which the words were spoken, "the Eagle has landed". (Armstrong, Neil, C.E. 1969)

Since the American Republic was the last of the <u>Chaos-Based</u> governments (under Lawful Law) to have started up in our modern-world, only to be quickly overthrown and replaced by an <u>Order-Based</u> government (under Legal law), this will be our last historical example, shown as proof. I will end this book and compilation of histories with a series of quotes given by the early American Revolutionaries, which displays the knowledge of both the Force of Order vs. the Force of Chaos in everyday life. Thanks for reading.

Quotes by early American Revolutionaries

Thomas Jefferson said that, "I wish it were possible to obtain a single amendment to our Constitution … I (would) deny their power of making paper money or anything else a legal tender". (Jefferson, Thomas C.E. 1798)

George Washington said, "government is not reason, it is not eloquence - it is force! Like fire it is a dangerous servant and a fearful master; never for a moment should it be left to irresponsible action". (Washington, George, C.E. 18th century)

Thomas Jefferson said that, "I sincerely believe, with you, that banking establishments are more dangerous than standing armies; and that the principle of spending money to be paid by posterity (debt), under the name of funding, is but swindling futurity on a large scale". (Jefferson, Thomas C.E. 1816)

Benjamin Franklin is reported to have been asked by a Mrs. Powel of Philadelphia exactly what kind of government has been created for the new United States of America and to this question Franklin replied, "A Republic, if you can keep it". (McHenry, James, C.E. 1787)

Thomas Jefferson said that, "it is incumbent on every generation to pay its own debts as it goes. A principle which if acted on would save one-half the wars of the world". (Jefferson, Thomas C.E. 1820)

James Wilson said that any, "government, in my humble opinion, should be formed to secure and to enlarge the exercise of the natural rights of its members, and every government, which has not done this in view, is not a government of the legitimate kind". (Wilson, James, C.E. 1790 - 1791)

Thomas Jefferson said that, "Bank paper must be suppressed, and the circulating medium must be restored to the nation to whom it belongs". (Jefferson, Thomas, C.E. 1813)

Samuel Adams said that, "if men of wisdom and knowledge, of moderation and temperance, of patience, fortitude and perseverance, of sobriety and with true republican simplicity of manners, of zeal for the honour of the Supreme Being and the welfare of the commonwealth; if men possessed of these other excellent qualities are chosen to fill the seats of government, we may expect that our affairs will rest on a solid and permanent foundation". (Adams, Samuel, C.E. 1780)

Noah Webster warned that, "before a standing army can rule, the people must be disarmed, as they are in almost every country in Europe. The supreme power in America cannot enforce unjust laws by the sword; because the whole body of the people are armed and constitute a force superior to any band of regular troops". (Webster, Noah, C.E. 1787)

WORKS CITED

<u>Acheson, Mel</u>, (C.E. 2013), "Round Sun vs. Foursquare theory", Published March 7, 2013 online website The Thunderbolts Project, A Voice for the Electric Universe", Reviewed and retrieved in 2018 from http://www.thunderbolts.info/wp/2013/03/07/round-sun-vs-foursquare-theory-2/

<u>Acheson, Mel</u>, (C.E. 2018), "Chapter 1 - What is the electric universe?", Published September 16, 2013 online website The Thunderbolts Project, A Voice for the Electric Universe" Reviewed and retrieved in 2018 from https://www.thunderbolts.info/wp/2013/09/16/chapter-1-beginners-guide/

<u>'Ali, 'Abdullah Yusuf, M.M. Pickthall</u>,(C.E. 1999), "The Meaning of the Glorious Qur'ān", Explanatory Translation by <u>Muhammad Marmaduke Pickthall, Dar Al-Kitab Al-Masri - Cairo, Dar Al-Kitab Allubnani</u> - Beirut. Published by Amana Publications, Beltsville, Maryland, USA.

Adams, Henry Brooks, (C.E. 1907), "The Education of Henry Brooks", Autobiographical and self-published in C.E. 1907, Boston, Massachusetts, USA. Commercial publication (after death) in C.E. 1918). Quotation found in Chapter 16.

Adams, Samuel, (C.E. 1780), Letter from Samuel Adams to Elbridge Gerry - written November 27, 1780, United States of America.

Armstrong, Neil, (C.E. 1969), Broadcast quotation from American astronaut Neil Armstrong through NASA Mission Control 4:18 PM, July 20, 1969 during Apollo 11 mission to the Lunar Surface.

Biers, Richard Lee, (C.E. 2011), "The Electron Myth", published November 2011 as an Amazon.com electronic eBook, Free Soil, Michigan, USA.

Black's Law Dictionary 10th edition, (C.E. 2014), West Group Editor Bryan A, Garner. ISBN-13: 978-0314613004. Originally composed in C.E. 1860 by Henry Campbell Black.

Boole, George, (C.E. 1854), "An Investigation of the Laws of Thought on Which are Founded the Mathematical Theories of Logic and Probabilities", originally published by Walton and Maberly, London, England, United Kingdom.

Bruce, C.E.R., (C.E. 2018) "Papers of Dr. Charles Edward Rhodes Bruce (1902-1979)", Archive Collection, Edinburgh University Library Special Collections, Reference GB 237 Coll-643. Archive Reference https://archiveshub.jisc.ac.uk/data/gb237-coll-643 Reviewed and retrieved in 2018 from https://archiveshub.jisc.ac.uk/search/archives/04dac8f0-4419-3b0d-9256-e38a8e5e5244

Caldwell, (Janet), Taylor, (C.E. 1965), "Pillar of Iron", historical fiction. Publisher Doubleday, New York City, New York, USA. Historical fiction based on Cicero, Marcus Tullius, Ancient Rome, Italy, approx. B.C.E. 55 using excerpts from his "Speech in Defense of Sestius" and summary of his political philosophy.

Einstein, Albert, (C.E. 1905), "Does the Inertia of a Body Depend Upon Its Energy Content?" - Fourth paper submitted and published by the journal Annalen Der Physik on September 27, C.E. 1905. Publisher: Wiley-VCH owned by John Wiley & Sons, Weinheim, Germany.

Farrell, Joseph P., (C.E. 2010), "Babylon's Bankers: The Alchemy of Deep Physics, High Finance and Ancient Religion", Publisher: Feral House, Port Townsend, WA, USA. ISBN 978-1-93259-579-6.

Freud, Sigmund, (C.E. 1923), "The Ego and the Id", Publisher: Internationaler Psychoanalytischer Verlag, Vienna, Austria, W.W. Norton & Company.

Gobry, Pascal-Emmanuel, (C.E. 2016), "Big Science is Broken", THE WEEK magazine online, published April 18, 2016. Reviewed and retrieved in 2017 from www.theweek.com/articles/618141/big-science-broken

Howard, Milford W., (C.E. 1895), "The American Plutocracy", Publisher Holland Publishing Company, New York, New York, USA.

Hossenfelder, Sabine, (C.E. 2018), Article entitled "Dear Dr B: Should I study string theory?" Posted on the online website BackRe(action)on May 11, 2018 under the label of Dear Dr B, Quantum Gravity. Reviewed and retrieved in 2018 from http://backreaction.blogspot.com/2018/05/dear-dr-b-should-i-study-string-theory.html

Jefferson, Thomas, (C.E. 1798), Letter from Thomas Jefferson to John Taylor - written 1798, United States of America. Found in "The papers of Thomas Jefferson. Retirement series" - 10:64. Publisher: Princeton University Press. Publication date(s) 2004 to 2016.

Jefferson, Thomas, (C.E. 1813), Letter from Thomas Jefferson to John Wayles Eppes - written September 11, 1813, United States of America. Found in "The papers of Thomas Jefferson. Retirement series" - 6:494. Publisher: Princeton University Press. Publication date(s) 2004 to 2016.

Jefferson, Thomas, (C.E. 1816), Letter from Thomas Jefferson to John Taylor - written May 28, 1816. Found in "The papers of Thomas Jefferson. Retirement series" - 10:89. Publisher: Princeton University Press. Publication date(s) 2004 to 2016.

Jefferson, Thomas, (C.E. 1820). Letter from Thomas Jefferson to Anthoine Louis Claude Destutt de Tracy - written December 26, 1820. Found in "The papers of Thomas Jefferson. Retirement series" - 10:175. Publisher: Princeton University Press. Publication date(s) 2004 to 2016.

Jung, C.G., (C.E. 1944), "Psychology and Alchemy" second edition C.E. 1968 Collected Works. London, Routledge, England.

Körtvélyessy, László (C.E. 1998),"The Electric Universe" ISBN 963-8243-19-8, EFO, Budapest.

Körtvélyessy, László (C.E. 2018), Quotation taken from the Electric Universe online Website. Retrieved in 2018 from http://www.the-electric-universe.info/the_book.html

MacDonald, Ronald; Rowen, Robert M.D., (C.E. 2009), "They Own It All (Including You!) By Means of Toxic Currency", www.booksurge.com.

Maltz, Maxwell M.D., F.I.C.S., (C.E. 1960), "Psycho-Cybernetics", Publisher Prentice-Hall Inc. Englewood Cliffs, New Jersey, USA.

McHenry, James, (C.E. 1787), Quotation by Benjamin Franklin written in the notes of Doctor James McHenry, delegate from Maryland to the Constitutional Convention of C.E. 1787. Published in The American Historical Review, Volume 11, C.E. 1906. Page 618. Publisher American Historical Association, Oxford University Press, Oxford, England, United Kingdom.

McCutcheon, Mark (C.E. 2010), "The Final Theory - Rethinking our Scientific Legacy", Publisher Universal Publishers, Boca Raton, Florida USA. Originally published C.E. 2002. ISBN 10:1-59942-866-0.

Plato, (B.C.E. 380), "The Republic", political philosophy written in ancient Greek language, one copy can be found in Paris, France, Bibliothéque Nationale (9th century).

Talbott, David; Thornhill, Wallace, (C.E. 2005), "Thunderbolts of the Gods, A Radical Reinterpretation of Human History and the Evolution of the Solar System", Mikamar Publishing, Portland, Oregon, USA.

Tewari, Paramhamsa, (C.E. 1996), "Spiritual Foundations", See online website www.tewari.org - Reviewed and retrieved in 2017 from that website.

The Holy Bible, King James Version. Cambridge Edition: C.E. 1769; King James Bible Online, C.E. 2017 source reviewed and retrieved from www.kingjamesbibleonline.org.

Torpedo, (C.E. 1883) from his Pamphlet "The Electric Universe: Flashing thoughts for consideration and facts from many sources". Pamphlet believed to be attributed to Sir Robert Stout using the pen-name of Torpedo. Pamphlet found in The Pamphlet Collection of Sir Robert Stout: Volume 86 entitled "A Motive Force in Nature" Auckland, New Zealand.

Velikovsky, Immanuel, (C.E. 1950), "Worlds in Collision", MacMillan Publishers, London, England, United Kingdom.

Washington, George, (C.E. 18th century), Quotation found in The Christian Science Journal, Volume 20, Number 8, November of C.E. 1902, "Liberty and Government" by W.M., Quote Page 465, Publisher: Christian Science Publishing Society, Boston, Massachusetts, USA.

Webster, Noah, (C.E. 1787), "An Examination of the Leading Principles of the Federal Constitution", Pamphlets 58 to 61 dated October 10, C.E. 1787 - Printed in "The Founders Constitution" Volume 1, Chapter 16, Document 17, Publisher: University of Chicago Press, Chicago, Illinois, USA.

Webster, Noah, (C.E. 1806), "A Compendious Dictionary of the English Language". Sidney's Press, New Haven, Connecticut, USA.

William of Ockham, (C.E. 14th century), source retrieved in 2017 from https://en.wikipedia.org/wiki/Occam%27s_razor. However, the phrase may be also be sourced to Libert Froidmont in his text "Philosophia Christiana de Anima" (C.E. 1649).

<u>Wilson, James</u>, (C.E. 1790 - 1791), "The Works of the Honorable James Wilson L.L.D. - Lectures on Law, delivered in the College of Philadelphia - Volume II", Publisher: Lorenzo Press, Philadelphia, Pennsylvania, USA. Published C.E. 1804. A series of lectures on the law that James Wilson later wrote down. Quote found on page 466.

<u>Zeep, Ira G.</u>, (C.E. 2000), "A Muslim primer: Beginner's guide to
Islam, Volume 2". Published by University of Arkansas Press, USA.

www.ingramcontent.com/pod-product-compliance
Lightning Source LLC
Chambersburg PA
CBHW031603210526
45464CB00004B/1413